Additional Praise for James W. Perkinson's *White Theology: Outing Supremacy in Modernity*

"In this fascinating and inspiring book, Perkinson accomplishes two things that seem virtually impossible. First, he makes visible the invisible. While black reality in the United States has been subject to countless analyses, white reality appears to be impossible to grasp. The book reads the latter in light of the former and covers new ground. Second, Perkinson rethinks the basis of transformation. Key are not intention and moral appeal, as is commonly assumed, but the alternative energies and powers that arise in the midst of repression. Rather than painting a bleak picture of life under pressure, the book finds passion and vitality in unexpected places. The result is a profound reshaping of white reality and theology that has the potential to set the stage for things to come."

— Joerg Rieger, Professor of Systematic Theology, Perkins School of Theology, Southern Methodist University

Black Religion / Womanist Thought / Social Justice
Series Editor Linda E. Thomas
Published by Palgrave

White Theology

Outing Supremacy in Modernity

By

James W. Perkinson

First published 2004 by
PALGRAVE MACMILLAN™
175 Fifth Avenue, New York, N.Y. 10010 and
Houndmills, Basingstoke, Hampshire, England RG21 6XS
Companies and representatives throughout the world.

PALGRAVE MACMILLAN is the global academic imprint of the Palgrave Macmillan division of St. Martin's Press, LLC and of Palgrave Macmillan Ltd. Macmillan® is a registered trademark in the United States, United Kingdom and other countries. Palgrave is a registered trademark in the European Union and other countries.

ISBN 1–4039–6583–8 (cloth)
ISBN 1–4039–6584–6 (pbk.)

Library of Congress Cataloging-in-Publication Data

Perkinson, James W.
  A white theology of solidarity : signified upon and sounded out / by James Perkinson.
    p. cm.—(Religion/culture/critique)
  Includes bibliographical references and index.
  ISBN 1–4039–6583–8 (cloth)—ISBN 1–4039–6584–6 (pbk.)
  1. Race relations—Religious aspects—Christianity. 2. Race relations—United States. I. Title. II. Series.

BT734.2.P47 2004
277.3′0089—dc22   2004046488

A catalogue record for this book is available from the British Library.

Design by Newgen Imaging Systems (P) Ltd., Chennai, India.

First edition: November 2004
10 9 8 7 6 5 4 3 2 1
Printed in the United States of America.

Contents

Acknowledgments

In one sense, this book has its beginnings on the basketball court of my youth. The outbreak of a sudden love affair with the game at the age of eight plunged me into the world of urban sport in the neighborhood I grew up in (which became majority black before my freshman year) and the inner city magnet school I attended from junior high through high school. There a passion to perform with abandon exploded into consciousness and whatever loves I have developed since (and there have been a number) owe their intensity to what I began to discover about both fear and desire in that world. The ancestry of this project goes back far behind its first conception in my mind.

The obvious place to start the praise song, of course, is my family: my parents, Ralph and Mildred Perkinson, who taught me to respect all and pay particular attention to the underdog; my brother Jerry, with whom I shared all of my earliest triumphs and traumas; my grandparents on one side who modeled staunch German industry and understatement (that somewhere along the way got lost in my own development); and my grandmother on the other whose cantankerous liveliness taught us all how not to die before we die.

Special friends all along the way also deserve mention: Deborah who introduced me to the ghetto and "called me out" with profound sensitivity and equally profound relentlessness; the entire family of Church of the Messiah, who counseled and confronted over 15 years, Champ and Magic and Mark and Gwen who schooled me in the neighborhood; Tapasananda at the ashram in Ganges, MI, who provided the supportive space to do much of the preliminary writing. My love of teaching I owe to Robert Werenski, who mentored and befriended far beyond expectation in early years of graduate work, whose eyes lit up worlds.

The University of Chicago opened to me as delight in dialogue about almost everything with Robin and Jim, Zolani and Jan, Matthew and Deborah and Linda, Gerald and Kazi. Fellow student Nahum Chandler's rich insights on W. E. B. Du Bois in the context of the African American Studies Workshop provoked some of the thoughts that eventually galvanized this writing. Professor David Tracy quickly saw my path through theory as distinctly my own and provided the license I needed to pull together insight from outside the discipline. He believed in me beyond my own confidence. Michael Eric Dyson became, for a few of us willing to straddle institutions, a model of analytical acumen and acrobatic passion that left an indelible stamp. From my very first American Academy of Religions presentation for the Black Theology Group in 1993, James Cone and Will Coleman have given unsought support and affirmation. When Dwight Hopkins and Katherine Tanner came on board with the Divinity School at the inception of my proposal process, their ready input and strong encouragement made for very easy writing and long hours of worry-free deliberation. For such handling by my committee, I am deeply grateful. Katherine has ever been ready with letters of support in seeking presentation and publication opportunities; Dwight is finally due thanks for initiating the contact with Palgrave and tendering the support that resulted in the book. Amanda Johnson, my editor at the press, has been constant in excitement from our first contact, and seen a book inside a huge sheaf of dissertation.

Over the years since graduation in 1997, others have entered the scene of my wandering ruminations: the entire faculty of Ecumenical Theological Seminary has always been quick with affirmation and interest; my department colleagues at Marygrove College likewise have been ever encouraging and accommodating of my needs. Oscar and Ken and George, thanks especially! Bill Wylie Kellerman close to home and Ched Myers across the country have been immeasurably influential as intimates in the struggle and conspirators in theory. Charles Long and the annual gathering of hard-thinking, freewheeling religionists at Farmington, Maine have graciously welcomed me into their deliberations for two years now, and contributed sharply to my perceptions of our modern dilemma.

My friends among the arts community are also present on these pages in cadence and color, rhythm and risk with language. Ron Allen especially has been inspiration and consolation, the granddaddy of Detroit street polyphony, breaking open parallel universes inside this one, inside me. To my own inner circle of support through the limbo years of adjunct work—Bev and Peg and Anna: I could not have made it without you!

And finally, and most crucially, my wife, Lily Mendoza, has believed in me and my work since the day she met me more than two years ago—often beyond my own belief in myself—and has been my deepest confidant and dearest refuge at every turn since.

Series Editors' Preface

Jim Perkinson's book stakes out a novel space in today's study of religion and theology. It takes on directly the challenge from African American intellectuals, and black theologians especially, to white male scholars regarding the notion of race. In these pages, you will find perhaps the most comprehensive discussion of "whiteness" by a white male theologian. Perkinson wields the poetic pen as he brings to bear an interdisciplinary analytical scalpel in the dissection of what is a "white race." Part autobiography (i.e., a white man living in a northern ghetto), part an aesthetic performance (i.e., Perkinson is a poet), and partly the result of a sharp and careful mind (i.e., the author earned his Ph.D. at the University of Chicago Divinity School), this book will keep you engaged throughout.

What is race? What is white supremacy from a theological perspective? For the author, power in America is disproportionately distributed among white citizens and thereby enables the latter the luxury to avoid defining and thinking about a "white" identity. Yet, for Perkinson, the American white racial contract establishes the norm for how all Americans perceive themselves and those around them.

Amidst contemporary conversations over policy, morality, business, the family, and culture, it is refreshing to hear a new and well-documented claim about the proverbial "elephant in the room." For Perkinson, the white elephant constantly avoided and not talked about is the white "race" in America. What does it mean rationally to be white? How does it feel to occupy the privileged position in a racial hierarchy? And what implications can one draw from the white community's access to and control over much power, property, and privilege? Perkinson gifts us with a theology of responsibility and wholeness, not just for the white community, but for all citizens on this space called earth.

Jim Perkinson's groundbreaking work continues the quality and breadth of publications in the black religion/womanist theology/social justice series. The series publishes both authored and edited manuscripts that have depth, breadth, and theoretical edge and addresses both academic and nonspecialist audiences. It will produce works engaging any dimension of black religion or womanist thought as they pertain to social justice. Womanist thought is a new approach in the study of African American women's perspectives. The series will include a variety of African American religious expressions. By this we mean traditions such as Protestant and Catholic Christianity, Islam, Judaism, Humanism, African diasporic practices, religion and gender, religion and black gays/lesbians, ecological justice issues, African American religiosity and its relation to African religions, new black religious movements (e.g., Daddy Grace, Father Divine, or the Nation of Islam), or religious dimensions in African American "secular" experiences (such as the spiritual aspects of aesthetic efforts like the Harlem Renaissance and literary giants such as James Baldwin, or the religious fervor of the Black Consciousness movement, or the religion of compassion in the black women's club movement).

Dwight N. Hopkins, University of Chicago Divinity School
Linda E. Thomas, Lutheran School of Theology at Chicago

Introduction

Imagine a graduate classroom in an old church building in a blighted neighborhood at the edge of Detroit's struggling downtown area in 1997. The class is a seminary course in Social Ethics composed of three female African American *savants*, four white male wanna-be pastors, and a white male professor. Two-thirds of the way through the course, the women suddenly "throw down." "We have now studied scripture from the point of view of liberated slaves and empowered peasants, we have examined Columbus' crusade from the perspective of native resister and African revolutionary, and looked at contemporary social movements of the gendered, the racialized and the queered. What do you guys think? Do you think white supremacy is still a problem today?" The classroom climate becomes instantly palpable. Its feel is electrified sweat. The women are poised like cats with tails twitching, listening through every pore. Every one of the four men splutters his response. None of the four can speak to the question; all speak away from it, in dissembling rambles about unrelated subjects.

In one sense, what follows in this book is an exegesis of that condition of shuffling in-articulation. White America is largely unconscious and mute, unable to address the question of its identity as white. Power normally does not have to give an account of its own basis of operation. Such an *apologia* would already amount to disclosure of the arbitrariness that is the ground of most prerogatives of domination. But here the mask had slipped anyway. And the men as well as the women knew it. The inability formally to account for whiteness in all of its continuing organization of privilege, power, and property found anxious expression in an "apologetics of tone" that was less than honest. It is that question of tone, of resonant silences and discordant ramblings, that is the very stuff of my interrogation in this writing (Spivak, 1992, 177–181).

More formally, this book is a response to the gauntlet thrown down by Black Theologian James Cone in the 1960s to a white mainstream Christianity that has gone largely unaddressed by white male scholars since then. The book is an attempt to hear and respond. It offers a critical and constructive articulation of the theological meaning of white racial supremacy in shaping institutional life, personal relationship, and cultural practice in America at the turn of the millennium. Though concentrating part of its analysis on theological texts, the argument here contends that cultural habit and social conditioning are equally as determinative for racial domination as overt discourses. The root that must be exposed is *underneath* the word even as it is underneath the skin, in the very bones and capillaries of American culture, like a metastasized cancer. The problem is not only what is said but also what is assumed.

I argue that the subjectivity produced by white supremacy remains "theological" even when it is secular. The task is not so much criticism of texts as it is "outing" a form of subjectivity that is common to much of the country at large. I address the whiteness of mainstream theology by way of the *theological-ness of mainstream whiteness*. The burden of the book is to reexamine white race privilege in relationship to its historical genesis as a modality of "lived theology," and its practical continuation as a habit of secular embodiment. The goal is to articulate a white theology of responsibility, responding to and in alliance with the claims of Cone (and others) that color is and must be near the core of any American projection of integrity or "salvation" (i.e., "wholeness") worthy of the name.

But it is also important to note that this book is not committed to theology *per se*, but chooses to think in the terms of that discourse because it specifies the particular kind of religious history and the particular mode of discursive potency that has leveraged racial trauma in this country. The book could also have been written as an exercise in articulating "white religion." But as my own background is itself Christian and as it is Christian supremacy that has given birth to white supremacy historically, "theology" more accurately designates the kind of power that must be combated than would a more encompassing term like "religion." The theology offered here is then not "normal" Christian theology but rather seeks to pursue its subject in deep engagement with multiple nontheological interlocutors. The task is exposure of white subjectivity before it appears as the subject of a discourse like theology, in that inchoate region where, as historian of religions Charles Long says, experience is first wrestled into category (Long, 1986, 8). The aim is to reimagine whiteness as a *critical cultural construct.*

In its *critical* elaborations, the work makes use of a multidisciplinary approach to the question of whiteness to underscore the deep effects of racial formation in the modern West and elaborates a tripartite challenge to white theologians, in particular, to learn how to confess and redress white power in their intellectual work. That theological challenge insists that the confrontation of white privilege must be entertained and sustained across the full range of human "being"—the large-scale structures of socioeconomic privilege, the face-to-face politics of personal power, and the more hidden and unconscious logics of cultural habit and erotic embodiment. The central claim of that confrontation focuses on soteriology, arguing that whiteness has functioned in modernity as a surrogate form of "salvation," a mythic presumption of wholeness.

In its *cultural* work, the book pursues an exploratory methodology of what might be called "hermeneutical reflexivity." The presumption here is that oppressed populations often know more about the power position (and modes of identification) of their oppressors than the oppressors do themselves. Such knowledge is part of the acquired survival skills of people who are forced to develop keen powers of perception and anticipation in order to avoid increased abuse. While cognizant of the constantly shifting terrain of racialized perception (mobilizing ethnic stereotypes in very different ways in different times and places and among different groups), the book concentrates its attention on the black/white divide as still representing the most fraught flash point for racial encounter historically in the country and constituting the most difficult site of racial confrontation for white people both externally and internally. Part of coming to consciousness of oneself as white, I then argue, involves daring to look into black eyes and not deny the reflection.

The first moment in elaborating a white theology is one of hearing and internalizing black critiques that are both overt and oblique. It entails confrontation with the embarrassment of having already been "found out" by one's (in this case) most frightening other. I am not suggesting that white self-knowledge is *only* to be gained from encounter with black culture and people—inevitably, different kinds of whiteness emerge in response to encounters with differently racialized groups. In this work, however, I am concerned primarily with that practice of whiteness that has historically projected blackness as its most incorrigible other. Paramount, for this posture of listening, will be the voice of W. E. B. Du Bois in trying to write the phenomenology of race without duplicating its binary perceptions.

Equally important for white conscientization, however, is black cultural production away from the scene of encounter. Black religious ritualization, black aesthetic innovation, and black political demonstration encode an ongoing interrogation of the depth structures of America that consistently finds "hearing" and embrace in the rest of the country. At core, such "soundings of" are also "significations upon" a meaning-structure that enters into white identity and practice as well. To the degree the country is a shared space, black creativity inevitably and irresistibly infiltrates as well as resists white cultural development. Learning, as a white, to sit before and be taught by such knowledge "from without" is finally also an exercise in discovering unsuspected possibilities within. Thus in one sense the book constitutes a *midrash* upon the (in)visibility of blackness inside the whiteness of America. Part of the task for white theology is to exhibit white identity as already beholden to and in part constituted by black cultural creativity.

The other work of the book is *constructive*. Having explored the demand for critical confession of whiteness as oppression for others, and opened up black culture as the most challenging space of encounter necessary for white people to begin to confront the meaning of their own white embodiment, the task is then one of delineating and re-envisioning whiteness. Here the formulation reiterates Du Bois's preoccupation with doubleness: for whites, it involves learning to confess complicity in the continuing organization of illicit benefit (at the expense of people of color) while simultaneously struggling against the presumptions of white privilege (as a choice, not just a condition). The possibility is a dance on a razor's edge—neither disavowing the history nor cooperating with the identity. For white theology, imagining a more self-conscious and less pretentious form of subjectivity will entail re-constructing its practices under the theological rubrics of exorcism, re-initiation (or baptism) and lifelong apostasy.

In actual organization, the book opens with autobiography and general survey and then offers a kind of chiastic structure. The section entitled *White Privilege and Black Power* interrelates the questions of personal change and political challenge in trying to "divine" a collective condition. Chapter 1 tells the story of my own struggle to become racially "conscious" precipitated by living in an eastside Detroit neighborhood for most of my adult years while chapter 2 surveys the fault lines of race that continue to rupture American life at the turn of the twenty-first century and the theological landscape that has emerged as a result.

The second and third sections mirror each other chiastically and develop a comparative consideration of racial identities by exploring the body in time and space. *History, Consciousness, and Performance* moves from tracing the career of modern white supremacy as a "lived soteriology" and black struggles to counter such (chapter 3) through a phenomenological reading of Du Bois's description of the violence of "being racialized" (chapter 4) to a brief exploration of black modes of displacing oppression via performance (chapter 5). *Presumption, Initiation, and Practice* works inversely and reflectively from that centripetal itinerary and moves from a consideration of specific practices of whiteness (chapter 6) through a consideration of the depth of the work necessary to challenge supremacy (chapter 7) to an outline of a different kind of (possible) white integrity (chapter 8).

Each chapter initiates its particular concerns with corresponding reflection upon urban struggle—especially in connection with the events surrounding the beating of Rodney King and the 1992 Los Angeles upheaval. In one sense, urban insurgence serves as a kind of leitmotif of the book—that moment when continuing injustice refuses complicity with continuing silence and erupts in a demand to be heard. Theology does not normally take its bearings from unrest. But given its genesis in a first-century Palestinian resistance movement whose public takeover of a major city center of its day (the Temple precincts in Jerusalem) precipitated the state reaction of arrest and execution that became that theology's galvanizing datum, perhaps it should. In any case, here it will. Undoing the muteness of white ways of being begins with learning to hear the urgency of black anguish. And part of the responsibility of making theological sense of an outraged sensibility is finding a style of writing that does not mask the meaning in its mode of theorizing. My text then weaves a certain element of poetic rhythm into its recitation—and this also by way of acknowledgment of its major interlocutor. In arguing that whiteness in America is already beholden to blackness in a profound manner, the writing itself seeks to reflect that debt.

1

White Privilege and Black Power

In the eighteenth and nineteenth centuries in America, black slaves absorbed, appropriated, and actually "stole," from their white masters a blondy-haired, blue-eyed Jesus of orthodox Christian doctrine who was "blackened" in their reconstruction of him not so much in feature as in figure, not so much in skin tone as in bodily bearing, not so much in content of preaching as in meaning. This "contraband Jesus" authorized for them, at great risk: an *outlaw space* (the Invisible Institution), a *criminal time* (night-songs segmenting the darkness into percussive rhythms considered insurrectionary), an *illegal identity* (of personhood, rather than "property")—and all of that in an *illegitimate* black Christian discourse whose specific difference from white Christian discourse was coded and carried in a necessarily surreptitious bodily style of profoundly critical significance.[1] Black practice cultivated ritual nuance—a God known in graceful glance, a Jesus loved in saving syncopation, a Spirit celebrated in potent possession. And those ritual instincts gave birth, finally, in the twentieth century, to a theology questioning not just the content, but also the form of godliness. James Cone became in the 1960s the best-known protagonist of the blackness of tongue giving theological expression to such a thickness of practice by rendering Black Power defiance and vision explicit in religious discourse.

The opening section of the work that here explores the meaning of such practice (and its discourse) for white identity and white theology in twenty-first-century America offers autobiography as "necessary confession" of my own position over time in relationship to black experience and Christian conviction and surveys the landscape of race *vis-à-vis* the code of theology at the beginning of the new millennium. White Privilege and Black Power represent quick glosses on some of the themes tracked in the book. "Race" as definitive of the central relationship organizing social wherewithal and political opportunity in the history of the country yields a structure of white domination and non-white subordination in general, but also a particular paradox of empty passivity on one side of the color line and potent struggle on the other. Understanding the complexity of the contemporary American multiculture requires serious excavation of this historical paradox in white and black. The terms of the exploration take their specificity from the tactics of the resistance already noted: we will labor to uncover the codes of race in the infrastructures of space, time, and the body in seeking to outline the meaning of whiteness for the new millennium.

1

White Boy in the Ghetto

God has chosen what is black in America to shame the whites. In a society where white is equated with good and black is defined as bad, humanity and divinity mean an unqualified identification with blackness.

—James Cone (God of the Oppressed, 225)

In late April of 1992, when Los Angeles erupted in the wake of the Simi Valley verdict of "innocent" for the white police officers who had been video-taped beating Rodney King in March of 1991, I was beginning the process of writing the dissertation that this book is based upon. The four days of upheaval signaled a country still in the throes of racial strife and political turmoil over its history. That history of white supremacy, black slavery, red genocide, brown migration, and yellow labor obviously remained fraught with contradiction and anguish, violence and reaction. With numerous friends at the University of Chicago, marching in the streets in protest of the verdict and organizing on campus to confront the currents of racism still operative there became my own particular response to the crisis Los Angeles represented. In many ways, Los Angeles was a smoke signal on the horizon. In some sense, this book is an attempt to decipher the still smoldering fires—a burning that did not begin with Los Angeles and will not end with this writing. The ghetto heat that seared itself on the eyeballs of an entire country one more time in 1992, over the course of my own 52 years of life, has increasingly fired the passion and consumed the thinking that I try to combine in my particular vocation as a religious studies scholar and activist/performance artist. Los Angeles reemphasized the urgency and Chicago offered the immediate context for action during my grad school years, but inner-city

Detroit has been my primary locus of learning. For more than 20 years before and after the time in Chicago, an eastside, hard-living neighborhood has served as both street-school and soul-scourge—a place at once home, haven, hell, and continuous harangue for a white-skinned male seeking to live responsibly in a country and world of profound tragedy and remarkable creativity.

Detroit Pedagogy

It is the trope of "tragedy" that looms largest in my mind whenever my white co-citizens pronounce the word "innocent" over their treatment of nonwhite others here and abroad. America has ever been tagged as an "epic" happening in world history in Euro-American telling. It emerges as a story of good versus evil, light against dark, might making right in the ascendance of England's sons to the heights of world dominance over the course of more than 300 years of taking land from natives and labor from blacks and love from women and resources from around the globe. A hydra-headed and still-living "ideology of superiority" licenses all of the taking as "fitting" and indeed as a "white man's burden" of responsibility to set the rest of the world right (invasion of Iraq in 2003 is simply the latest in a centuries-long litany). This epic tale of America is the "deep code" of white being all the way up to the present moment. But it is "tragedy" that is the right reckoning by any measure from the shadow side of events. People of color in general, women in intimate quarters, and blacks in their very bodies and being, have been forced, by the violence brought to bear, to know "America" as a tragic reality, with all such means in terms not only of suffering imposed from without, but of the inner compromises necessary to survive a brutal condition. America, for them, is a tale of innocence plundered, choices made between competing evils, and wisdom baked in a harsh oven. My own tale in the mix is one of a gradual weaning from desperate clinging to an "epic identity of innocence" and initiation into a more complex and contested self-understanding inculcated over years at the hands and under the tongues of those who know America from underneath.

I grew up in a neighborhood in transition in Cincinnati, Ohio and learned the hard way that skin color matters. As the youngest son in the only Protestant family on a block otherwise teeming with Catholic kids, I had few friends among the parochial school cadre that controlled the lone basketball court behind our houses. The day after I brought a black public school buddy onto the gravel court at an off hour to shoot a few hoops, I was beaten by the other kids for breaking a code I did not even know existed. The court was "white space." Later, during the summer between my sophomore and junior years of high school, I would likewise be beaten by black males for daring to talk with a black female friend on the playground of my former elementary school. The year was 1967, the cities had already erupted, and those particular school grounds had become "black space." There, too, I was breaking code. But the consequence for the body of crossing such a boundary of the mind was not limited only to violence. In the inner-city high school I attended, I also learned to play basketball with a concern for finesse and an instinct for contact that was stylistically "urban" and personally great fun. That double experience of pain and pleasure has remained the single constant of my wrestling with the question of race ever since.

By the time I graduated from the University of Cincinnati in 1974 with a degree in business management, I knew that my real interest was not business, but something more like theology and the arts. The move to Detroit that year was precipitated by a deep thirst for things communal and charismatic—I had been more profoundly shaped by the student prayer group I helped lead than by the classes I attended, and knew only that the quality of inner experience opened by that spirit had, somehow, to be expressed in a shared lifestyle. A small gathering of assorted folk from around the country in an eastside Detroit Episcopal parish—there to share life and longing, assets and aspirations, gift and groan and gaffe, in a bold struggle to make sense of faith in the fraught terrain of 1970s North America—caught my attention through a mutual friend, and I visited, and reacted, and joined.

Church of the Messiah was host structure—a carcass of a church, disemboweled by the latest wave of white flight following the 1967 Detroit rebellion—that in early 1974 was being re-fashioned as a communal experiment in pooling resources, sharing living quarters, collaborating in decision-making, mobilizing energy to "minister to" a devastated neighborhood. The new little community organized its living in various "extended family" households within a five block radius of the church, putting as many as 20 people under one roof over here, 12 over there, 7 in yet another brick home just around the corner. The commitment was a poverty-level lifestyle—everything in common and available for whatever need turned up in the neighborhood. Black and white, married and single, adult and child labored, and laughed, and fought, and fumbled for more than 15 years to reinvent "church" on a model more intimate and resonant with the Beatitudes, more radical and responsive to the local politics, more visionary and vibrant than the typical American take on the gospel as a promise of prosperity. By far, life in this little communal experiment became the deepest part of my education—not least for the "plunge" into ghetto reality it entailed.

But my "baptism" into such a reality began naïve and presumptuous. For most of us who were white in the group, the motive for years was "mission"—"doing good" in the slum for the sake of spiritual solace and egoistic pride. Only gradually did some of our eyes open to the actuality. Inner city neighbors, for the most part, were far more equipped to make meaning out of the madness, to create in the midst of the chaos, than those of us "doing ministry." The teenagers next door would convene dance floors complete with broomstick "mikes" and plastic bat "guitars," lamp-shade strobe lights and a boom-box band in a crowded living room, and do imaginary battle with their favorite MCs and DJs. Younger kids converted the abandoned auto in the chicory-choked lot outside my window into a bi-level platform for gymnastic flips and flops and wild floatings onto the tattered box-springs positioned by the front tires. Older women could relate riveting stories for hours on their crumbling front stoops, while older men chopped rusting Corvairs and Camaros into cavalier hybrids in the back.

Only over years did I finally grasp that I was not there so much to give as to grow, to participate in a reciprocity of gifts and needs that I did not start, could not control, and would not end. My particular ability had primarily to do with the access to assets and expertise located elsewhere in the social order that my white skin conferred on me as privilege. I could help bring such to bear in the neighborhood in ways that might contribute to some amount of benefit for the neighbors. But they commanded the survival skills: the humor; the ability to celebrate things as simple as breathing each day or as complex as Coleman Young getting elected to another term as mayor; the quickness to share food or bus fare or care for children; the young persons' flair for hip-hop as a stop gap in the flood of despair—performing, like their blues-crooning parents before them, existential "judo" on the jive of life as a racialized minority in the midst of the American Dream. I learned. Gradually. At cost to the patience of my neighbors. And I discovered that I was indeed white. But also that I had a choice to be more than just that!

Over time, my tastes changed and my passions found new depths. From a narcissistic belief that a born-again experience was the one and only non-heretical interpretation of life, I shifted to a growing recognition of spiritedness and vitality everywhere. These latter showed up in unlikely packages and unseemly places: the schizophrenic Vietnam war vet named Roland whose "word salad" conversations with invisible companions on the street corner next to my house might suddenly be interrupted, in an aside to me, with a depth of insight into my own secret struggles, and a brazenness of confrontation of such, that would take my breath away. The quirky grandmother—who ate and chattered incessantly because of her constant worrying about everything from the street wanderings of a young transvestite friend to the latest gun control initiative on the ballot, from the food delivery package that had not yet arrived from the shut-in ministry to the Karposi's carcinoma that her AIDS-infected son was succumbing to—whose small apartment nonetheless became the neighborhood house of hospitality, taking in whoever, for whatever reason, showed up homeless on her doorstep. The 30-something, single mother, whose election as the first housing-coop board president, in the building the church had rehabbed and helped organize into a resident-owned enterprise, presented her with a crisis: she had never led anything before, but emerged as a savvy and shrewd convener of a struggling new business venture. But these neighborhood manifestations of irrepressible spirit were not the only source of unanticipated inspiration.

The world at large increasingly divulged its own wily wonders to an eye pried open to the unusual. For me, the change was not primarily a matter of belief; I simply *saw* the thing in front of me and could not deny the potency. Meetings with Islamic imams and Jewish rabbis in a Jewish, Muslim, Christian tri-logue disclosed shrewd insights about justice. Involvement with a Hindu ashram on the west side of Michigan during summer retreats plunged me into disciplined meditation and rich musing on the relationship between inner attention and outer effect. Concern for questions of economic oppression led into sustained engagement with innovators of cooperative housing experiments and worker-owned businesses, Latin American base community advocates (on a trip to Nicaragua), and rural Mississippi community development entrepreneurs (in a program of exchange between youth from Detroit and young people from Mendenhall). The name invoked in these varied projects might be Christ or it might be Marx, the claim embodied might be faith or it might be liberation; the effect in every case, was vital human commitment and remarkable spiritual ferment. And all of it demanded ardent appreciation. My own "passion to know" began to shift away from straight theology toward a growing fascination with "cultures of contestation"—leading to loves ranging from mysticism to montage, zen koans to *Vodun* rhythms, Tibetan mandalas to Celtic shamans.

But around it all—underneath it, beside it, in front of it—was the world outside my doorstep, throbbing with beat, yelling against the heat, streetwise and strapped down. A black world that awakened untold depths of joy in me—and irresolvable anguish at the complexity of trying to live responsibility against the grain of white power! It is that world that, for more than 20 years now, has most confronted me with myself, most challenged me to be more than myself, most embraced whatever nascent offerings of syncopated rhyme, stutter-stepped dance-mime, or ribald story-line I could manage in honor of the rich humanity around me. Graduate studies at the University of Chicago became largely a foray in wrapping multi-disciplinary theory around that ongoing experience on eastside Detroit to become as polyglot as possible in addressing the world of white supremacy about the urgency of the hour and the depth of the dilemma. What the city taught me can be summarized under the rubrics of perfidy, quandary, and liberty.

White Perfidy

Eastside anguish revealed, under Chicago analysis, an American generality. Whiteness exists in this country today as a color-blind fiction of innocence, publicly posturing itself as the neutral pursuit of the Dream, wishing well on all sides, intending equality, sorry for poverty, certain of the uprightness of its own vision of ascent into the gated bliss of sole proprietorship (Harris, 1). What it intends to own, without evil intent, is simply the whole earth. I am speaking in explicitly abstract terms. There is no one literally white. The term is a cipher for a social position of domination underwritten by a text of absolution. Actual individual persons of pallor may or may not cooperate with the prerogatives their skin tones leverage (Leki, 16; Perkinson, 1997, 202, 206). But if they choose *not* to cooperate, that effort indeed *requires* effort and exacts costs (Cone, 1975, 97, 101, 242). Merely to "live, move and have one's being" on the contemporary landscape without protest, is, for those of us who look vaguely like me, to be complicit. What we are complicit with is a (now globalizing) regime that continues to mobilize and impose the markers of darkness for the sake of an accumulation of resources and rights on the light side of the racial divide. But that divide is largely invisible today from inside the institutional domains of white pleasantness.

On its other side, however, the invisibility casts a shadow that is irreducibly physical. Policing forces, jail bars, mall surveillances, sports stadium gazes from on high, store clerk stares from the end of the aisle, helicopter search lights, dollar bill extractions from the wallet in excess of prices charged elsewhere, disease treatments more likely to severe limbs than prescribe pills, and so on—all the indices of the continuing difference darkness "makes" have their real purchase on the body, not the imagination. A simple sample: the average black male dies in this land at the age of 64.3 years old—often of stress-related diseases—*before* receiving back a single dollar in social security contributions (Parker, 15). The body is buried; but its lifetime of labored contributions goes where? The social security system functions as a huge transfer payment to the more insulated and less troubled that is not at all innocent of race.

The analysis holds for most of the institutional processes in our country. Add up the effects of real estate "black balling" and insurance redlining, black and white pay differentials for the same work (75 cents on the dollar on average), asset disparities for similar households (white median wealth that is ten times greater than black), mortgage lending practices that reward white skin with twice the credit-tab on average as black skin, differences in employment and/or promotional likelihood for equal skills, expenditure per pupil in school systems euphemistically coded as "suburban" and "urban," likelihood of arrest, conviction, incarceration, and indeed execution if young, black, and male compared with white counterparts, the "shadow-banking" (alternative) credit industry that steers more than $300 billion per year out of impoverished urban neighborhoods of color and into (if the paper trail is followed all the way "up") Fortune 500 companies—and the "cost" of melanin for its inheritors is astronomical. "Blackness" is the condition of an ongoing squeeze play. What is usually not also indicated when the tab is tallied is that the benefits largely accumulate where the sky and the skin are fairer. My white wherewithal is *constituted in* Afro- (as indeed Latino- and Filipino- and aboriginal-, etc.) American impoverishment. Their loss is my gain. The relationship is utterly asymmetrical and the asymmetry utterly relational.

White Quandary

Set in such a context, the classroom dilemma described in the opening paragraph is far deeper than the irritations of a merely dilatory response. It is symptomatic. The white males themselves were from varied backgrounds—a Scandinavian-gened Lutheran pastor, a slightly bull-doggish Germanic Episcopal-priest hopeful, a shy, sincere rural-dwelling youngster struggling with depression, and a former Vietnam vet, homeless-shelter-convert envisioning ministry as a lifetime of work with other homeless men north of the city. For each, in quite different ways, seminary study was a stretch, a demand to engage texts and tighten arguments that imposed new disciplines of mind and new twists to the tongue. But the academy as "academy" was nonetheless familiar cultural turf—the sentence structure in the classroom, the presuppositions about prerogatives of speech, the impersonal character of written authority, the bureaucratic power of professorial position were not new. That pedagogical *ethos* is historically European. Foucault's ideas of institutional genealogies and discursive archaeologies would disclose here a relatively common universe of cultural habituation and prosecution of power shaping the soul of the seminary and the psyche of the student alike in this case (Foucault, 1980, 1977, 1970; Welch, 1985; Hopkins, 1997).

Not so the women. New York anthropologist Thomas Kochmann already in the 1970s detailed the difference cultural background makes in the salt-and-pepper classroom in a book entitled *Black and White Styles in Conflict*. "In conflict," and "conflicting," indeed! The significance of the difference is in the subtlety of the style and the style of the subtlety. Black character in this country tends to reflect a pedagogical shaping in the home and on the street, in the barber and beauty shop, on the basketball court or in the hall that is often distinctive for its personalism (Kochman, 18, 21, 24; Bynum, 86–88). What constitutes knowledge, what qualifies as authority, what makes up argument, what governs process, what adjudicates persuasion, what amplifies eloquence, what satisfies the demand to adhere is all conditioned quite differently in most black and white communities (Kochman, 19–20, 24, 34, 26, 29, 40). The reports on the specificity of black pedagogy and the potency of black creativity are manifold: Henry Louis Gates, Jr.'s "Signifying," Houston Baker's "Blues Banter," Susan B. Willis' "Specifying," Zora Neal Hurston's "Lying," Geneva Smitherman's "Talk," Edward Bynum's uncovering of an ongoing articulation of "Ashe," Theophus Smith's notion of "Conjure," Toni Morrison's erudite "Funk," Cornel West's uplift of "Antiphonal Kinesthesis," Amanda Posterfield's "Incantation," Trish Rose's "Black Noise," all the rappin' and raggin', jukin' and jivin', dozens-playing and doggin' out that points to the utter thickness and sheer prodigality of black cultural innovation.[1]

These ongoing elaborations of rhythmic potency also mark out the terrain of what must be anticipated and allowed to live in any seminary pedagogy seeking to probe a presence of divinity in history as Word or Wisdom. Those four white guys in my classroom that day didn't stand a chance of remaining un-dissected, un-deconstructed, un-dissed, or un-devilishly put on (Perkinson, 1994, 200–203). The sad thing was they had no words by which to learn from the dismantling their guts told them they were undergoing in that classroom encounter. The women, on the other hand, knew the precise place of vulnerability and were quite adept at launching small sonic probes, little vibratory vectors of interrogation—sometimes as silent as a shared roll of the eyes or a mere grunt of punctuation—that convened instant community among themselves in the presence of the rest of us. Remarkably, it was also the women who "rescued" the men when the anxiety underneath their own faltering inarticulation grew palpable to the point of embarrassment. One of the sisters would jump in with a comment or question that took the pressure off by directing the attention away.

The women *knew*—and knew *that* they knew—in a way that the men could not (Foucault, 1980, 81–86). It is that epistemological "prevenience" (like in "prevenient grace") that is the point of the approach this book takes. Therein, I would argue, lies a vernacular advent of spirit that is the hallmark of the tradition a Christian seminary is committed to trying to teach. The Judeo-Christian tradition is a tradition that began in the cry of Exodus slaves (Ex. 2: 23–25), prioritized its own identity in reference to the cry of widows, orphans, resident aliens, and the poor in its legal codes (Ex. 22: 21–27), privileged the prophetic uptake of that cry when its own leadership became repressive (Jer. 8: 18–9: 6; Amos 4: 1–2, 8: 4–6), found a paradigmatic expression of those premises in one who championed the cause of the oppressed in his own day until he himself ended up embodying that cry (Mk. 15: 33–39) on an instrument of capital punishment called a cross (Perkinson, 2001, 105–106, 113). The Judeo-Christian tradition is preeminently a tradition that asserts that divinity shows up in history where humanity is most under pressure and least comforted, most constrained and least resourced (Perkinson, 1998, 166–168). It is a tradition of reading "God" present in the quietest moan of the marginalized, the slightest sigh of the suppressed, even the tiniest groan emerging from the ground of being itself (according to Paul in Romans 8: 18–26). Deity, here, is not associated so much with dominating majesty as with irrepressible dignity and dissent in even its most diminutive forms (Ezek. 9: 4). The venue is not wide-open sky, but a crack in the sidewalk, opportunistically seized by blues-singing chicory.

Educating for recognition of *that* deity is a matter of opening the ear and eye to the entire undercurrent of eloquent "soundings" of unjust order.[2] Which was the very thing happening in that classroom through the subtle dynamics of a racialized confrontation! Blackness, in that moment, was the mood of indigo improvisation on ignorance. It was the triple-time dance around the halt and lame tongue of white maleness, the discomfiting of arrogance, the ironic volatilization of privilege and passive presumption. The shame of the moment was that the men hardly knew "Spirit" had suddenly appeared. They were trained to look for large recognizable-ness, white skin, order, theological tameness, and textual propriety. Their sought-after Subject had surprised by coming forward in the key of syncopated slyness, slithering and supple as a wave of sea, uncontainable, laughing in fractal proliferations of meaning working that room at the speed of light, revelation like a tiny orgasm of a knowledge they didn't have. Unfortunately—Christian seminaries in the main are *constituted* in that kind of "agnosticism" in this country. They don't know. This book is one small attempt to counter the incomprehension.

White Liberty

And finally it is also necessary to note up front something that will be labored into theory in the discussion that follows.

Since returning to Detroit in 1996 to live in the same neighborhood and teach in a local college and seminary, I have become a regular participant in the urban poetry scene. The particular artists I hang out with combine the influences of jazz, funk, and hip-hop and work in the margins between spoken word's emphasis on performance and a more literary emphasis on artistry. They are largely African American by ancestry and inner city by choice. My own choice to locate myself in this cadre rather than other possibilities has largely to do with a peculiar kind of "infection"—I have grown to love eastside intensity and Detroit urbanity and they have gotten inside of my own psyche. In my poetry in particular (and my writing in general) this takes the form of a fascination with syncopated rhythm and sharp-edged percussion. I do not claim these affinities as innate leanings, but more as the sound of the neighborhood reproducing itself in my own sensibilities after 20 years. I have been altered in my basic "life-beat." Trying to live apart from an environment that resonates to some of these same frequencies would now be almost unthinkable. The response of some of my friends—white and black—has been occasionally to tell me I have "gone black." And herein lies the dilemma that animates and motivates this book.

Eminem, Kid Rock, and the Beastie Boys all to the good, I nevertheless think it is absurd for white people to pretend that in contemporary America they can somehow "be black" in a kind of inversion of the phenomenon of "passing." The very attempt is delusional and part of the problem of white ignorance of the social and political realities of white racial power. At the same time, I think it is imperative for white people to be profoundly and consciously changed—both inwardly and outwardly—by blackness. What could this possibly mean?

James Cone's very first book entitled *Black Theology and Black Power* allows (if barely) for the possibility that people who look like me might become "black [persons] in white skins" through their political commitment (Cone, 1969, 3). I would never claim such for myself nor, in my own hearing, allow other whites to make a similar claim (about themselves or about me) without challenge. Whenever I have received such an appellation from black friends, however, I have been at once deeply touched, thoroughly humbled, and not a little frightened. The fear, for me, is an index of responsibility. Such honorary inclusion in blackness is a profoundly generous gesture but also an unwarranted affirmation. In my estimation, only a *lifetime* of unrelenting struggle against the ideology of white superiority and the materiality of white control (struggle "for," i.e., redistribution of the power, property, and privilege that white skin has assembled for itself historically in this modern world) could ever qualify one for embrace as an ally. I have not yet proved worthy. I have taken some risks, but have not so far dared push the challenge to the point of becoming a physical target. Nor would doing so imply the right to think of myself as somehow having magically changed my skin color.

I do, however, want to challenge myself and other whites to engage such a lifetime struggle and to recognize that doing so entails work at the level not only of institutional change and personal openness but of cultural conditioning. I do not believe the working of white supremacy inside of white people can be successfully combated without some measure of recognition of black creativity. And genuine recognition means both appreciation and incorporation. When I encounter powerful expressions of other people's humanity, I do not merely say to myself "that's nice" and walk on. I study their finesse and beauty and learn. I take in the potency and let some of its influence alter my own sense of identity. Staying alive spiritually until one actually dies physically, for me, means continuing to change and experiment and stretch, even as a mature adult. It implies a refusal to lock into a "comfort zone" of certainty about identity that subtly erodes the capacity to create new meaning or relate to a different way of being.

For me, the pilgrimage out of white cultural conditioning and into something more resonant with certain strands of black expression has been long, painful, and profoundly rewarding. Years of gradually increasing awareness of my own whiteness and growing appreciation for black cultural resistance have been punctuated with moments of breakthrough. At a party in a nearby apartment in the late 1970s, I suddenly stopped trying to imitate my neighbors on the dance floor, and dared give license to the beat in my own bones, and was quickly called out by the others around me: "Look at Perkinson! Party over here! A white man with moves!" I never forgot the immediacy of that affirmation and the liberty it created in my own memory. On an April afternoon in the mid-1980s, I witnessed Charles Adams of Hartford Memorial Baptist Church fame—a whooper of legendary proportions in black preaching circles—whip up a largely white political rally protesting U.S. Central American foreign policy into a frenzy of call-response in something under three minutes, and experienced a reaction of recognition within. The next morning I was lay-preacher at Messiah Church and incorporated something of rhythmic cadence in my own ten-minute parlance for the first time. The racially mixed congregation rose up in spontaneous shouting and applause! Later exposure to fellow Detroiter and renowned philosopher Michael Eric Dyson's style of classroom teaching—beginning slow and measured and then mounting into a caterwauling concatenation of imagery and theory that left the head spinning and the body heaving in laughter at the sheer excess of genius—also left its mark. I began incorporating syncopated poetics into my own academics, wrapping "rational argument" in "hip-hop entrainment," making "modality" itself carry some of the weight of the meaning. The style that emerged is my own and has gained a certain notoriety in academic circles. But its nurturance into expression is entirely the gift of the culture around me. I have no rights of ownership, but owe a deep debt of gratitude whose most fitting reciprocation is committed struggle for greater racial justice. This book is one small gesture of thanks.

2

The Crisis of Race in the New Millennium

It is not accurate or, for these times, bold enough to just say that America has a race problem ; . . .America is a race problem.

—Haki R. Madhubuti *(Why L.A. Happened: Implications of the '92 Los Angeles Rebellion*, xiii)

On May 4, 1992—twenty-three years to the day after the burning of cities during the 1960s had issued in a black reparations demand being delivered to the white liberal congregation of Riverside Church in New York—the last flames of South Central, Los Angeles were put out, ending four days of unrest. Between New York in the late 1960s and LA in the early 90s, an entire nation had managed to go nowhere in its grasp of its own tragic history. A week after the LA event, a popular national magazine featured the uprising in fold-out glossies. On the overleaf were two Asian-looking women consoling one another before a burning house as backdrop. On the facing page, a police car teetered on edge under the onslaught of four pairs of bulging tan biceps, coached by a set of black hands to the right. And at bottom, a powerful left-handed swing of an iron pole shattered storefront glass, wielded by a lightly pigmented cowboy-booted someone with fury in his mind and wind in his wavy hair. Meanwhile the brief blurb explaining the photos read, in effect, "blacks take revenge!" With almost no black skin in sight!

Two pages further on, the visual dissonance was even more damning of this written text. A broken storefront was depicted disgorging a multicultural mélange of activity. In evidence were three

African-looking, nine Latino-looking, two Anglo-looking, and one Asian-looking opportunists, all busy redeploying commodities in a "feasting" at once materially opportunistic and ritually carnivalesque. It was a gesture as much of the economically threatened as of the racially angered. But it had already been misread as "black planet rising."[1] That simple mis-cognition in the media—graphically offering popular opinion its own presuppositions—is the very signature of white supremacy.

The Crisis of Race in Current Social Practice

After centuries of operation, white supremacy in action in America has yet to be fully "outed" and exorcised. While the clarity of the reparations foray in 1969 highlighted the economic stakes of the struggle against such a supremacy, the symbolic density of "LA-in-May" represents something of its social complexity and its future responsibility. The blackness that popular media mis-saw in the theater of American color, is, historically, the prototypical perceptual object of race in this country, deriving from the time of slavery. Typically (but not exclusively), it is whiteness that sees that blackness, and in so doing, seeks to reassure itself (usually unconsciously) of its status. This "seeing" is the first moment of supremacy. The denigration (discrimination, destruction) that often follows in its wake—such as the shooting of Latasha Harlens[2] or the beating of Rodney King or even the derogatory reporting of the rioting—is supremacy in action. At this point in our history, the seeing and the acting constitute a social system and a cultural coding that affects and infects virtually everyone in the country. Even a Korean shopkeeper shooting a black teenager, or a Latino policeman beating a black male adult, are implicated in an order that finds its primary reference in white privilege and its favorite scapegoat in black disadvantage. But where denigrated blackness (the shooting of Harlens, the beating of King) indeed marked the *root* grievance of the 1992 uprising, it is the multicultural complexity of the disturbance that marks the new *range* of significance of that "blackness." Race is no simple matter in this country, even when its effect is a simplistic perception.

Yet for all that, something must also be said on the other side. The magazine certainly mis-spoke its seeing; but within that mis-speaking, behind and underneath it, is a truth. It is a truth that, for all the Watts and Newarks, all the Detroits and Philadelphias, in the 1960s, indeed, all the uprisings and riots, the little daily revolts and the occasional large-scale rebellions over more than 300 years, has yet to find an adequate hearing in the white public sphere. It is the truth of the terror of black living in America since its inception, and the temerity of white ignorance out of which it is produced. Los Angeles in 1992 is part of a tradition of denial and its discontents that must indeed be recognized in light of its forerunners even if it is responded to differently. It remains my conviction here that black skin—in the fact of the difference it has been *made* to bear—has endured a brutality and bedevilment quite unmatched in the experience of any other ethnic group since the slaughter of the original inhabitants of this country. In response, it has brought up from its depths, a creativity also unmatched. The double constitution of blackness as a history of pain and its overcoming remains as yet unintegrated in the cultural identity and social structure of the nation at large and thus demands a hearing alongside of every other crisis of identity troubling the political consciousness of the country. And whiteness remains *its* crisis. Such is the historical premise out of which I write. Such is the crisis I want to be responsible to.

Pretext: The Continuing Crisis of Color

The multicultural subject of the popular culture text finds its root in the history of *black* struggle, even if its range is dispersed through a growing spectrum of racial difference. That root was given a most cogent Christian witness in James Cone's 1969 publication of *Black Theology and Black Power* that began unpacking the public meaning of black struggle in the key of theology. While numerous social scientific scholars[3] recognize the identity politics of the Black Power movement as representing a new moment in race relations in the long history of white supremacist control of this country, it is Cone's work that spells out its most incorrigible meaning as an intimation of a different kind of "salvation." The exposé that follows this chapter takes its motive force from Cone's claim. Black Power, for all its youthful excess and sexist arrogance, nonetheless broke through a web of silence and a demand for sameness (on the part of white culture), to articulate a counterdemand of respect for difference. Urban confrontation became its most notorious mode of expression. In the spectrum of civil rights activity from early bus boycott to lunch-counter sit-in, from freedom rides to voter registrations, it is not until Black Panthers appear on the streets of Oakland "strapped down" (carrying weapons in plain view) that the rawest form of "racial nightmare" for white supremacy materializes most graphically. Armed black anger carried an irrepressible message. Of course (!) the response of the dominant (white) culture was unremitting repression in the immediate moment and relentless rollback of affirmative action gains over the course of ensuing decades. But Cone's articulation most provocatively names the continuing theological stakes. Black Power as a social movement found its extreme of expressive urgency in urban fire. Cone claims that fire as "christological" and projects its only possible future "banking" as a new realization of American wholeness. Los Angeles in 1992 was a rejoinder of continuing confrontation, one more time, screaming black pain as warning and black empowerment as its necessary remedy. What had changed from the 1960s was that now that cry of refusal echoed anguish among Mexican migrants and Salvadoran refugees, Korean capitalists and Jamaican rastas, Filipino war vets and even welfare-dependent Anglos. As the black freedom movement had offered encouragement to all manner of (other) ethnicized and gendered struggle (the feminist movement, the American Indian Movement, the United Farm Workers, the Puerto Rican independence initiative, etc.), so in Los Angeles black rage crystallized a "rainbow riot" of outrage. How such might be understood as a soteriological demand aimed at white power is the subject of this book.

But I get ahead of myself here. If the public eruption of the crisis I explore serves well as opening, it is the more hidden and ongoing everyday practices of that crisis that must exercise our theoretical vigilance in what follows. And here we do not have to look very far to find our subject. Even the academic study of religion, seemingly so distant from the raw struggle of the streets, exhibits the effects of that crisis in its expectations. For it is a crisis that is also a continuity.

Text: The Everyday Crisis of Race

It remains possible in the American Academy of Religions annual meetings in recent years to attend sessions hosted by groups carrying names like "Black Theology Group," "Afro-American Religious History Group," "Hispanic American Religion, Culture and Society Group," "Native Traditions in the Americas Group," and "Womanist Approaches to Religion and Society Group." In recent years, it has also become possible to attend sessions in a track called "Men's Studies in Religion." But it has not yet been possible to attend any session hosted by a "White Theology Group," or a "Euro-American Religious

History Group," or an "Anglo Approaches to Culture and Society Group." It is likewise possible to go into campus bookstores in most major cities in the United States and ask where the section on Black Literature is, or African American poetry, or Black Theology, without a single eyebrow being raised (though a request for Hispanic or Native or Womanist literatures may or may not turn up separate sections). But only in a few places has the nascent discipline of "Whiteness Studies" begun to make a place for itself and even then remains largely unheard of (much less "believed in" as legitimate) in society at large. The difference in these possibilities and responses witnesses to a prosaic but profound phenomenon in this culture.

In this particular moment of its history, knowledge is organized and produced in the academy (as well as almost everywhere else in American society) in a manner that presupposes various taken-for-granted processes of making distinctions (which remain quite matter-of-fact notwithstanding the notoriety of our much ballyhooed "culture wars"). Race and ethnicity (not to mention gender and class and sexual orientation) continue to be *differentially marked* in our institutional and interpersonal life together in this country. Addressing that phenomenon from a theological point of view that is self-confessedly "white" involves a peculiar work of "defamiliarizing." It means giving public display to commonsense categories such that their hidden strangeness and harrowing consequences are thrown into relief.

The everyday designations of difference ("black," "Latino," "native," etc.) just alluded to have, in fact, emerged historically in this country as the traces of something that is much more frightening and insidious than a merely prosaic "difference-making." They are haunted by the memory of a terror as original to "America" as the Puritan errand-into-the-wilderness[4] in the Seventeenth century, and as recurrent as the "prototypical" beating of Rodney King (with all of its aftermath) in the spring of 1992. The markings of otherness are symptomatic not just of an ideology of difference, but finally also a practice of violence. They point to a power of differentiation in connection with which people have been willing to steal and exploit, evict and imprison, beat and kill. They exhibit, that is, not only the (seemingly) irrepressible classificatory impulse of *racialization*, but also simultaneously the brute and brutal fact of *racism*. And this latter is largely the prerogative of those who have the institutional power to enforce their difference-making on others, for the sake of economic advantage, political power, and cultural self-congratulation. In America, racialization is practiced by (and upon) virtually everyone—black, white,

Latino, Filipino, Eskimo, Arab, and so on. Thus far in American history, however, racism, to the degree it references not just prejudice, but institutionalized practices of power (able to translate prejudice into advantage for one's own group at the expense of others), is largely white only.[5]

Theologically the signs of these practices of race beg to be read as "signs of the times" in the United States. They designate something profound about the American *Zeitgeist*—the spirit of the age "ghosting" the deepest wellsprings of our thought, bedeviling the innermost intentions of our action.[6] On a mundane physical level, they must be comprehended as integrally linked to our notions of land, property, and the human body, anthropologically to our sense of identity and community, ethically to our vision of meaning and purpose, metaphysically to our immediate horizon of finitude and our ultimate hope of wholeness. As our history attests, they indeed signify on a level where people struggle for their chosen lifestyles, discern moral values, calculate their chances for happiness, and decide questions of life and death. All of which is to say that in America the signs of racio-cultural difference intersect with what could be called the *soteriological level of existence* (delineated in more detail in chapter 3). Despite their quotidian manifestations, they finally have to do with ultimate concerns. They raise questions of "sin" and "salvation."

The Crisis in Contemporary Theological Practice

James Cone was the first theologian, black or white, to attempt to bring the interconnection of race and salvation into theoretical relationship to American theological production in general. Arguably, Black Theology has been extant in the Americas since the first slave conversion and has addressed itself to the conundrum of race in voices both clear and clairvoyant. But until Cone's 1969 integration of SNCC's "Black Power" cry into a theology that was explicitly "black" and "proud," the question of the meaning of Christian salvation at the level of racialization in America at large remained untheorized. In Cone's writing, what vernacular Black Theology had known from the beginning, and coded in various practices of (necessary) subterfuge and surreption, was given public voice in the very forms of "white" theological literariness. Under his pen, Black Theology took upon itself a proper name, moved into the theological mainstream, and demanded a hearing from white and non-white alike. But even then, it was not until Cone's 1975 work *God of the Oppressed* that the deep soteriological significance of blackness began to be rendered theologically explicit in his exposition of the Black Christ.

It is perhaps in this historic assertion that in America today *Christ is and must be racially black* that the profound question of the historical interrelationship of race and soteriology finds its most compelling formulation (Cone, 1975, 134). For many whites and no small number of blacks, the claim is simply shocking, a *non sequitur* of incarnational thinking that has provoked intense charges of ideologizing the gospel. The idea itself is not new. Marcus Garvey in particular gave it trenchant expression in the movement he launched in the early part of this century. But it was in the voice of Cone that black messianism first gained theological notoriety and it was in *God of the Oppressed* that it first attained theoretical sophistication.

The Crisis "as" Black Theology: Salvation and Counter-Racialization

In this latter work, Cone asserted that Black Theology's unequivocal insistence on blackness as a "christological title" in America today was a *theological*, not just a psychological or political, claim. For Cone, the representation of Christ as black was not to be understood as simply a kind of "ideological antidote," or anti-type, to white theology's own ideological captivity to a white (or supposedly colorless!) Jesus. It was not just a dictate arising from the psychological and political needs of black people for a savior "who looks like us." Rather, it was to be grasped as "the soteriological meaning of the particularity of [Jesus'] Jewishness" in the present context of white racism (1975, 134–135). "He *is* black," said Cone, "because he *was* a Jew" (1975, 134).

Cone validated blackness theologically by means of a dialectical understanding of "cross and resurrection" in relationship to the concepts of "particularity" and "universality." The cross represented for him "the particularity of divine suffering *in Israel's place* [as the 'Suffering Servant']" (1975, 135; emphasis added). And the resurrection signaled "God's conquest of oppression and injustice, disclosing that the divine freedom revealed in Israel's history is now available *to all*" (1975, 134; emphasis added). In effect, Cone insisted that the universalization inherent in the resurrection-triumph portends the actualization of God's presence in every particular "cross" of historical suffering that *any* marginalized people are made to undergo. It does not erase, but rather frees up *for other concrete contexts*, the particularity of God's identification with the Palestinian poor in the Jewishness of Jesus.

For Cone, this "real and not docetic" identification of the Risen Christ with the oppressed poor necessarily meant that in the context of white racism Jesus must be black. It could not be otherwise if the particular/universal dialectic of the cross and resurrection was taken seriously. Identification with explicitly Jewish suffering in first-century Palestine pointed irresistibly toward identification with black suffering in contemporary America. Saying such, however, did not imply for Cone that blackness might not become inappropriate as a christological title in some future moment or different context (indeed, admitting the relativity of such a title to time and place only squared it with other such context-specific titles, like the Roman-imperial "Son of God" and the Jewish-nationalist "Son of David"). But it did mean that here and now, "the blackness of Jesus brings out the soteriological meaning of his Jewishness for our contemporary situation" (1975, 134). It was "not simply a statement about skin color, but rather, the transcendental affirmation that God has not ever, no not ever, left the oppressed alone in struggle" (1975, 137).

What we have, in this theological conscription of blackness into service as a christological title, is a postcolonial reversal of the colonial use of soteriology to decide and structure the meaning of race (as we shall examine in chapter 3). Clearly here, the determining power of the respective discourses is turned around. In Cone's christology, race begins to have determining effects for the meaning of salvation. Racial signifiers are transvalued into the very grammar of an inverted ultimacy.

But something else is also happening in this move that has remained undertheorized to date. The reversal begun here is actually double. Not only has the structure of the relationship between salvation and racialization been reversed, but also its hierarchy of proximity to grace has been overturned. Or said the other way around, the presumption of soteriological precariousness has switched sides. *Contra* earlier Puritan notions, in Cone's discourse (in particular, as in Black and Womanist theologies in general), the greatest uncertainty about salvation must now be predicated *not of dark skin, but of light*. (Although Cone is careful to clarify that blackness is not itself a marker of innocence, but rather of the kind of condition a God of the oppressed chooses to occupy with incarnate flesh and spirit: black people can still sin and betray their calling, but black culture and suffering, on the current landscape of race, are a privileged place of encounter with the God of the least.) In soteriological considerations, it is the white side of the ledger of race that must be exhibited as the most troubling and troubled liability.

Cone himself hints at this at a number of points (indeed, some of them are more than just hints: they are pointed demarcations!). But his own particular focus in his work has been on a theological exegesis of blackness. He speaks to and for black people (albeit expressly in the hearing of whites) in a manner that subjects white theological discourse not only to black interests but, increasingly, to black discursive protocols, as well. He is not programmatically concerned about white destiny in the above reversal; it only comes up in his work tangentially. When he does address the (in his experience, rare) possibility of a positive white response to the thrust of his challenge, he tends to speak in terms of those white people having "become black" (Cone, 1975, 241; 1969, 3). It is not Cone's purpose to try to work out the soteriological meaning of blackness as a christological title *for white people in their whiteness.*[7] Indeed, he would probably resist any suggestion that he do so as yet one more attempt to appropriate black creativity for white benefit. (Cone has already been roundly criticized, within black religious scholarship, for implicitly reinforcing white theological norms too much as it is.) But the question remains in the air. If Cone's reconstruction of blackness as a christological title is taken seriously in its claim to be *theologically*—and not just politically or psychologically— motivated, what is its soteriological meaning for white people? Can white people be saved by a black Christ?[8] Must they? If so, how?

Putting the question in this latter form is already an abuse of the claim, however. The theology involved here does not so much concern itself with hermetically sealed contexts of salvation addressing themselves respectively to black and white people. Rather, "blackness," as a cultural construct, is being rewritten as a theological quality. It is not just black color, but black culture, *as a social poiesis,* that is being loaded with theological import. Such a move necessarily also loads white culture with theological import.

The broadest form of the question is simply: what does whiteness have to do with blackness in America today? Theologically, what is the relationship between them? Cone's claim, in its simplest articulation, is that blackness is christological and as such is salvific. The implication that arises from that claim and serves as the basic thesis for this entire project is that Cone's christological assertion of blackness as salvific in America specifies a sociocultural "site" that is integral to the salvation of *all* Americans, *including whites*. The exact meaning of the term "site" and the quality of relationship implied by the term "integral" will become clear only in the course of developing the subsequent chapters.

The Crisis Multiplied: Salvation in Multicultural Mélange

Since the emergence of the Black Power movement in the 1960s and its theological expression in Black Theology, numerous other theologies of color and difference have staked out claims on various terrains of oppression and struggle. Native American theology began offering its word at the same time as Cone's first publication in Vine Deloria's *Custer Died For Your Sins: An Indian Manifesto* and later intensified its challenge with *We Talk, You Listen* and *God is Red* and a host of riveting critiques like Robert Warrior's "Canaanites, Cowboys and Indians: Deliverance, Conquest, and Liberation." Latin American theology gained the high ground of alternative theorizing with Gustavo Gutierrez's *A Theology of Liberation* and emerged as a veritable discipline in its own right. Feminist theology quickly gave theological purchase to the issue of gender with the likes of Mary Daly's *Beyond God the Father: Towards a Philosophy of Women's Liberation* and Rosemary Ruether's *Sexism and God Talk: Toward A Critical Feminist Theology* and a plethora of prophetic- and post- and anti- Christian challenges that ranged widely, debated deeply, and theorized with profound insight.[9] Asian theologies like those of C. S. Song and Aloyius Pieris have brought sharp questions of interreligious encounter and conflict to the table and insisted on popular culture and folk religion as idioms not simply of false consciousness but of theological engagement and potential liberation. Womanist and Mujerista theologies have asserted their own differences of experience and analysis and sharply challenged and immensely enriched the reading of scripture and street alike in works like Delores Williams's *Sisters in the Wilderness: The Challenge of Womanist God-Talk* and Ada Marie Isasi-Diaz's *Mujerista Theology: A Theology for the 21st Century*. Asian American women like Hisako Kinukawa and Pui-lan Kowak have contributed yet other unique perspectives to the mix. More recently, the seismic thud of theological soundings coming from alternative sexualities has further complicated the concerns and plunged theology into the question of identity *de profundis* (i.e., Richard

Cleaver's *Know My Name: A Gay Liberation Theology*). In all of the ferment, Cone's solicitation of salvation as one of the defining thresholds of the debate continues to demand delineation. In such a postmodern mélange of struggle and stridency, of meanings being made and unmade from multiple vortices of pain and power, how can the possibility and responsibility of human wholeness be imagined? How can something like integrity be striven for or trust exercised in a soteriology?

At one very deep level the question can be organized as a dilemma of identity that is perhaps new in its historical urgency if not in its complexity. In a shrunken world order—for the foreseeable future presided over by American imperial power in service of transnational economic interest—twenty-first-century globalization presents traditional structures of belief and identification with a crisis of destabilization. Relentless urbanization of world populations fragments group filiation even as it coalesces bodies in ever smaller spaces. Communities crunched together in slum streets or represented together in the virtual beats of video, radio, and TV cameo (even while separated by police and concrete) increasingly borrow and beg from each other. At the level of youth culture, hip-hop has emerged from the hard pavement and harder encampments of the Bronx to become a kind of new world-idiom of adolescence-on-the-rise against its despised youthfulness. Recent protests like those witnessed in Seattle or Genoa over WTO manipulations of world trade in favor of already dominant parties throws up the prospect of a carnivalesque parade of concern: the rap generation in league with Third World farmers, blue-collar laborers marching with Green Peace conspirators, Catholic Workers making cause with cyberpunk hackers, anarcho-primitivist visionaries and mainstream environmentalist organizers sharing water bottles, indigenous rights activists and soccer moms arm-in-arm. The Los Angeles tumult at the turn of the last decade was merely a brief augury. But the reflection of such a riot of representation inside theological articulation makes for quite tortured expression of something like a "saved soul."

Serious theory recognizes that many if not most of us in the developed world now inhabit multiple, competing identities, worked out in ever-morphing modalities of belonging to numerous different affinity groups, none of which entirely defines our existential "position." We are now creatures of many collectivities, only one of which may offer promise of something like redemption in the traditional sense, but every one of which has bearing on our bodily integrity and substantial identity. I am indeed *constituted* (in my very bodily substance and life circumstance!) by my socioeconomic dependence upon a global metabolism of goods and services and signs (that privileges my place on the planet at the expense of most other places): by connections to Brazilian peasants through the beef industry (if I eat meat), to teenagers stitching Nikes in sweatshops in Taiwan or mothers making baseballs in Haiti (if I play or pay for sports), to the fate of rubber tappers in rain forests (if I drive a car with Michelins or Firestones), the straits of enslaved child pickers of cocoa in Africa (if I drink latte or eat candy), the baiting of the Latin American unemployed to trust life and limb to trafficking "coyotes" (to come north to "nirvana" and pick my fruits). I reside in and move through a succession of buildings whose materials are the matter and labor of countless multitudes stretched across an entire globe: could I hear the walls talk they would tell stories of great lament and hardship that my daily "deafness" only reinforces. More consciously I circulate among one or more family groupings, serial if not coexistent work settings, various recreational, therapeutic, artistic, and spiritual affiliations. And my sense of personal style and preferences for public presentation of my body in clothes, cosmetics, and accoutrements cobbles together an unlikely potpourri of creativities assembled by the advertising industry from a world stage forced to offer its sensibilities to the market. What could it possibly mean to claim a modicum of wholeness in the middle of such?

Out of the bricolage, theology has waxed heteroglossic, ferreting out the fire of resistance in every nuance of struggle and oppression, linking each local site of suffering to its global coefficient in colonial domination, patriarchal suppression, white subordination, capitalist exploitation, or heterosexist silencing. In the mix, Cone's claim of blackness as christologic inside of the American world hegemony begs—and receives—radical relativizing. Theologies of struggle come together in forums like the American Academy of Religions and listen to each other's brilliance, nod agreement about the excrement of the global system of transnational domination, but so far have not really forged a popular theology capable of articulating those multiple responses (to quite different experiences of suffering) in a single voice. And perhaps that is a good thing. But it also conduces to the continuing hegemony of white supremacy operating in concert with globalizing capital, while the disparate communities and constituencies struggling against such a domination compete for allegiances and assets.

Where theologies of struggle have not quite managed a common *logos*, however, it is interesting that the contradictions of youth subculture have responded to a common *rhythmos*. That hip-hop culture should emerge in the midst of contemporary globalization as a nearly ubiquitous idiom for youth seeking to grasp their own impossible production as "limbo people" is telling. Caught between the certainties of childhood leisure and adult labor, bombarded daily with the eroticized message to "consume" and yet enjoined by every adult institution overseeing their position to contain the hormonal frenzy thus aroused, young people urban and rural alike, black, white, Chicano, and Czech, find expressive outlet for their conflict in the percussive intensities of rap. Certainly this is in part a production of transnational marketing, but it is also arguably a recognition of something primordial. The beat is Afro-urban, the encoded pain American black, the defiance a latent insurgence. Perhaps the appeal is even genetic, tapping into a long ago repressed knowledge that the entire species is African in origin. While lacking concrete political organization, the energy, like any carnivalesque exhibition of discontent, is potentially incendiary. Given a congruent pedagogy, it could be mobilized as "global movement."

But the memory of "insurgent otherness" that the syncopation codifies is at the same time significant of a momentarily "eschatological" kind of creativity. It is able to invert the signs of its own domination into an alternative economy of pleasure. Rap's early appropriation of the vinyl record as an instrument of creative production in techniques of "scratching," for instance, gerrymandered a commodity designed for passive consumption into service of active contestation in a delightfully unforeseen manner. Though it hardly constituted revolutionary challenge and was quickly re-commodified in big label marketing strategies, rap DJ-ing—along with a host of other hip-hop tactics small and large (tagging privately owned space with public graffiti art, converting public street into private dance hall in ad hoc break dancing contests, inverting prison garb into fashion statement)—marked a redeployment of postindustrial desperation that simultaneously highlighted the continuing deafness of those outside. As African American literary critic Houston Baker asserted immediately after the Los Angeles upheaval, the condition for the possibility of really grasping what had happened in the Los Angeles debacle rested with the capacity to give ear to rap music: the deep base reverberations of the "urban black prophets of postmodernity" "sounding out" their circumstance in the tones of Grandmaster Flash's "Message" for almost two decades prior to the explosion (Baker, 1993, 45–48). If rap's "national anthem of the black inner city" could not achieve a competent "hearing," certainly no officially commissioned post mortem was going to illuminate the causes in the monotones of academic and bureaucratic policy-speak (Baker, 48).

True to its historical form as functionally "white," mainstream theology hardly registered a hiccup. Even in relationship to such a tiny gesture of resistant recoding as "scratching," it remains largely illiterate. Attempting to ask about the soteriological significance of a minority community's brief repositioning of one of the fetish items of domination (the vinyl record as commodity) in the typical seminary or divinity school would likely draw a dis-believing grunt of "huh"? Yet the modality of rap DJ-ing, as cipher for an entire global subculture of significant meaning-making in the face of crumbling family systems and eroding neighborhood networks, is certainly as much of a "sign of the times" as any biblical text that could be quoted in the context of the wildly efflorescing Pentecostal movement that is also going global in our day. But modern theology—at least in its mainstream protocols of production—has remained so immured in "the text" as to be almost unable to conceive of a question of salvation in a medium of rhythm. And this despite its own beginnings in an inarticulate groan (Ex. 2: 23–25), its own tradition of salvation by as slight a means as a "sigh" (Ezek. 9: 4–6), its prototypical promulgation in oral preaching (Lk. 4: 16–30) and ritual acting (Lk. 22: 7–23)! The relative "babel" of theological frames and claims regularly launched skyward at AAR is a mere monotone compared to the polyphonic cacophony of cultural refrains being innovated to resist and reframe the domain of globalizing forces on the ground in local communities around the world. The quest for some semblance of human wholeness and significant meaning in a world of absurdly chaotic violence is prolific and prodigal and not by any means exhausted by explicit "God-talk." But surely such "secular" activity is as bound up with the divine drive toward freedom that aches inchoately under the surface of all creation (as Paul so provocatively envisions in his letter to the Romans of his day) as any more typically "Christian" practice seeking to imitate or secure wholeness now and in the hereafter. At least such is the tact taken here and more explicitly explored in chapter 3.

In any event, the issue for this text is not that of the necessity for translation and alliance between disparate theologies of struggle and cultures of contestation, but of the outing, naming, and taming of white power in its covert coercions under the surface of a globalizing Western culture. In our time, the centripetal core of that global force field of power is imperial America, striding the planet like a rogue cowboy, guns in hand, tobacco in cheek—as sleek and sophisticated as any dry Texas drunk after a six-month bender! And part of the problem is that white supremacy as a social force *is* sophisticated and sleek in its most damaging modes of efficiency. In effect, it functions as a secretive pact of propertied white males, in the exposé of philosopher Charles Mills, that underwrites the more public social contract vaunted in business schools and nation states as the cornerstone of Western world–hegemony in the modern era (Mills, 4–7). The premonition that white is "right, bright and best" that haunts the perception of even many black children in this country has a root that is irreducibly religious and irredeemably vicious. It is the color-coding of a pathological historical terror repressed under a conviction of being superior that has driven the West into its pretension of being preeminent World Conqueror and Developer of the indigenous into full stature as human.

In the middle of that modern historical development, James Cone asserts that blackness is critical to wholeness in contemporary America. Whether or not a similar claim could be made for other experiences of racialized oppression is beyond the province of this particular work but certainly commands attention. Other dimensions of the experience of socialized suffering such as those associated with "womanhood," the "bent back" or abused body of poverty, or the "fascination for the same" supposedly definitive of homosexuality, are likewise dialectically "necessary" for the constituting of dominant forms of identity in imperial America and demand equally probing theorization as lying at the core of the question of integrity in this culture. Deep ecology and bioregionalism are raising a similar question about the constitution of human being with respect to local ecology. But here the dimension in question will remain race and the concentration will maintain its focus on the peculiar cast of features roughly identified as "African" that does seem to divine and disturb what is most frightening in the ferment of white cultural makeup as "white."

How Cone's claim should be negotiated by other populations of color in the space of America remains an open question that only those communities can settle. And indeed a reciprocal challenge inevitably accrues to communities colored "black" in relationship to the peculiarities of Amerindian, Hawaiian, Chinese, Latino, Filipino, *mestizo*, and *criollo* histories of suffering and the way those histories find voice in various theological and cultural calls for embrace and redress. Historically, however, it is black and red pain and protest that have most integrally and immediately entered into the material substance and cultural discourse of this country's emergence in the modern world. And it is these two relations of interdependence that remain in some ways most definitive for the strange cultural affinity operating as white identity. While brown history and struggle are integral to the history of the western part of the country and quite clearly will dominate the future of racialized struggle across the whole of America, and "olive" religious reaction to Western conquest and consumption defines contemporary geopolitical calculation, and Asian power looms on the horizon as the coming world-competitor, it is stolen native land and stolen black labor that anchor the *historical* genesis of white supremacist practice and discourse and continue to influence white response to other groups of "others." The way native resistance and persistence have factored into and influenced white identity and wherewithal is an entire project unto itself and thus here it will be enough to elaborate white soteriology in reference to the peculiarities of black ways of being.

Also beyond the boundaries of this particular work is the necessity for a serious probing of the roots of white supremacy in male fantasy and patriarchy. While I occasionally offer analysis of the erotics of race that have so profoundly structured its practice in society, limits of space prohibit developing the analysis the way it deserves. Focused examination of issues such as the way gender is constructed and deployed at the boundary of racialized difference, the way white male desire and fear are structured in relationship to projections about female bodies both white and dark, the role of homophobia in leveraging the cult-violence of castration in lynching rituals, the organization of institutional life in this country that manages white male fear of black maleness by way of the prison-industrial complex (incarcerating the black penis "away" from the white vagina), the terror in white male psyches that, were the playing field really to be leveled, white women might well prefer black males to themselves—all of the deep economy of terror and titillation that marks the negotiation of sexualities "normal" and "transnormal" in our society will have to await a future effort. Here it will be enough to begin to outline a white response to the history of black struggle that begins to answer to Black Theology's soteriological claims *about the meaning of race.*

Cone's claim has also been challenged from within African American scholarship on numerous fronts, not least of which is Victor Anderson's charge that Cone has invoked an ontologized notion of blackness in a manner that leads to a theoretical dead end. If blackness is given purchase within theological thought as definitive of the experience of people of African heritage in the Americas, and as constituted in either suffering or rebellion against the white supremacy that created the idea of blackness in the first place, then blackness itself becomes oppressive for blacks both in policing a certain kind of in-group conformity and in binding itself to whiteness as its ground of being and its ultimate concern (Anderson, 91). Anderson's critique is nuanced in indicating that his concern is "not to negate, but displace, decenter and transcend the determinative transactions and practices of ontological blackness over black life and experience" (Anderson, 17). His alternative is that of an aesthetic of the grotesque that licenses and liberates creative energies of intra-black dissent (among womanists, gays, lesbians, mixed race persons, etc.) and ebullient lifestyles of flourishing beyond the binary constraints of race.

And in one sense, my own project is an attempt to take Anderson's challenge seriously in relationship to whiteness and white people. At the deepest level, race, as a mark of difference between groups of human beings, does not exist. Bio-geneticists and anthropologists alike concur. It is merely a myth of the imagination, a social construct, a kind of perceptual shorthand to organize the visual field of bodies in the moment of encounter. But as a mere construct, it has obviously leveraged very real and troubling historical effects. Learning how to straddle that divide between mythic image and material effect—at once owning and struggling against the history of violent oppression white identity has unleashed on people of color and at the same time resisting the temptation to settle into a stereotypic set of meanings of white identity for oneself—is the task of integrity this work seeks to underwrite. Whiteness, too, needs to become "grotesque" in Anderson's sense of the term, an identity both "on trial" and "at play" in the encounter with other identities of culture and color. However, Cone's claim, in my estimation, continues to play a critical role in relationship to white supremacy. It throws down a kind of "reverse ontological" gauntlet at the feet of white people that must not be dismissed until whiteness itself ceases to function socially as a practical ontology and the actual social circumstances between racialized groups have been changed. The task is tricky: naming, exposing, denouncing, and "exorcising" the substantial (ontological?) effects of white racism in history while simultaneously deconstructing the very pretension to (ontological) reality that is the idolatrousness of the supremacy in the first place.

The Crisis in White Theology: Salvation as Ecumenical Dialogue?

Among white theologians in general (and certainly among white males in particular), Dutch journalist-theologian Theo Witvliet has offered the most sustained and in-depth response to the claims of Black Theology's explicitly black christology. While there is much that is of immense value in Witvliet's text, there is also a telling lacuna. Witvliet elaborately stakes out the coordinates of a profound challenge to white theology and at the same time leaves unaddressed, in his own theorizing, the deep meaning of the whiteness that Black Theology so profoundly interrogates.

Of particular import for the project here is Witvliet's careful delineation of the field of encounter (between whites and blacks) in terms of the difference Black Theology represents for the theology "we already know" (Witvliet, 1987, 3). This latter species—styling itself as "just theology"—must actually be understood as "white theology." According to Witvliet, assessments of the significance of Black Theology by white theologians are *already* complicated by tandem responses of guilt and superiority—responses that yet one more time function ideologically to exclude black culture and experience (Witvliet, 1987, 4). White silence on the history of slavery stands revealed here as no mere oversight, but rather as a form of interested avoidance absolutely central to the (white) mythology of whiteness. Any white intention to move beyond that mythology immediately faces a painful self-confrontation.

Witvliet further identifies two major obstacles for whites who would step into "the circle of black discourse" (Witvliet, 1987, 5). The sheer heterogeneous profusion of black texts on the one hand, and the vast "experience gulf" between white and black worlds on the other, complicates the move immensely. Of particular concern is the conundrum of difference that the (latter) "gap between worlds" represents for white theorists. The world of black experience, of racial oppression and the struggle against it, must necessarily be comprehended as irremediably "other" if it is not (yet one more time) to be reductively annexed to white experience and distorted by white theory. Such an otherness is itself instructive for a white mind-set that constantly seeks to secure its life-world as homogenous (or at least in control of any threat posed by difference). But it is also dangerously productive of its equally pernicious opposite. Dismissal of Black Theology as a "closed hermeneutic circle" of difference offering no point of contact or entry for white Christians obviously halts exchange before it can even begin. It simply serves to reinforce white gestures of exclusivity. This double-edged dilemma poses a profound question of hermeneutics.

For Witvliet, Black Theology's significance is universal and its import irresistible. Engagement with it is not some adventitious exercise of exoticism. Although itself subject to (internal) validation solely on the basis of its own articulated claims as a *liberation* theology, Black Theology's *hermeneutical* challenge is comprehensive in its theological scope. As an aspiration toward the deliverance of literally everyone on the American scene of race (and not just blacks), it cannot be dismissed simply as (a locally circumscribed) "survival" theology. Indeed, the very depth and breadth of its liberative dynamism drives it into a posture of concern and confrontation that is global in import—and thus demands global attention and response.[10]

Witvliet specifies the space of this global encounter as ecumenical. Ecumenism, for him, is the theological locus and practical place where contextual particularity can and must be translated into universal significance. But white theology is not Black Theology's priority interlocutor. Only after Black Theology first attends to the demands of a global, "intra-black ecumenism," and secondarily, to the need for a "peripheral ecumenism" involving other globally marginalized populations of color (e.g., Latin Americans, Asians, and minorities at home), does the question of an ecumenical encounter with white theology arise. For Witvliet, this third level of ecumenical interaction should take shape as a particular form of "polemics," seeking to expose the entire American situation as a colossus of contradiction affecting everyone, no matter his or her location or cultural formation. And here lies warrant for a very particular kind of rhetorical intervention.

Black *anger*, in Witvliet's estimation, is a critical modality of ecumenical integrity that must be embraced, by whites, as a form of pedagogy. In the encounter, it serves a threefold theological function of exorcism, exposé, and exhibition. On the one hand, black mobilization of indignation facilitates a self-liberating purgation of negative input. At the same time, it "checks" the claim often made by whites that white racism has been eradicated in well-intentioned individuals by "exploding" the systemic pathology of fear hiding underneath the claim. Alongside such existential effects, black fury also renders graphic the theological truth that the "unity" offered in Jesus Christ can only be comprehended as paradox. The hope for ultimate wholeness presented on the cross is a form of violated integrity. The way forward for any of us begins with confession of our complicity in that violation. Black anger, expressed within the crucible of ecumenical engagement, makes visible and urgent the complicity that white theology must learn to confess in America.

The Crisis in White Subjectivity: Salvation as Self-Conscious Discourse

The problem in this proposal lies in what remains unsaid. In his preface situating *Black Messiah* as a complement to his first book, *A Place in the Sun* (surveying various "Third World" liberation theologies), Witvliet indicates "both works are to be regarded as exercises in dealing with liberation theology *from another context*" (Witvliet, xvi; emphasis added). *Black Messiah* specifically seeks to "reflect on the significance of 'contextual' theology by studying *one particular* example of it" (xvi; emphasis added). Absent in such an advertisement, however, is self-critical attention to the eye that observes. In a work otherwise dense with erudition and empathy, there is no explicit delineation of Witvliet's own context as a white Dutch theologian in relationship to his long look at a contextual theology "from elsewhere." Such a project of "studying one particular example" of a contextual theology is misleading precisely from the point of view of contextual theology. Witvliet's own theological evaluation of Black Theology must itself be grasped as contextual and thus inevitably present as another context against which his one particular example is construed positively or negatively. If contextual theology is taken seriously in its claims while it is studied as "example"—and thus "object"—it necessarily complicates that studying process by requiring some measure of foregrounding of the (surreptitious) presence of a second contextual space from which the inquiry is carried out in the first place. To study one inevitably involves the presence of two. The lacuna the author exhibits here is characteristically "white."

Presumably Witvliet would answer that he has highlighted his context in the chapter (briefly sketched above) where he outlines both the history and the possibility of a white engagement with Black Theology in various local, national, and international precincts of ecumenism. And indeed, such a context may well be the only possible context for an intentional European engagement with Black Theology. But it is a context constructed somewhat off to the side of everyday life, an arena of encounter already circumscribed by particular ideological constraints and entry requirements. Witvliet's own *everyday* context as "white" and "Dutch" is *not* coterminous with that ecumenical arena. Whatever significance of whiteness pervades (consciously or unconsciously) his own sense of self is not derivative from that very occasional scene of encounter, but is built-in, at some level, to the less conscious and more habituated processes of identity-construction that already inform the meaning of "Europe." It is this place of personal habit that must be examined and theorized in relationship to the scrutiny directed toward Black Theology.

What is needed is not merely evaluation *of* Black Theology, but the articulation of a *white* theology that is itself contextual and thus accountable *to* the situation of race. In America, nothing less will suffice. The American dilemma that Black Theology articulates is difficult precisely in its complex doubleness. Its racial problematic (in an Althusserian sense) is a constant reproduction of blackness and whiteness in relationship to each other as the absolutes of an opposition (notwithstanding the way other color-coded groups are also made to occupy the position of "other" by white supremacy). The meaning of either term is necessarily parasitic upon the other. They operate together. Thus, at one very profound level, Black Theology does not constitute another context, but rather the self-same context out of which any self-consciously white theology would have to speak if it were to take itself seriously as white. Yes, Black Theology represents a profound difference *from* white theology; it is demonstrably and irretrievably other. But that very otherness is also simultaneously constitutive *of* whiteness, in various modes of white appropriation and stereotyping.

From a perspective inside of America, racialization appears as a convoluted circulation of identity and difference that is *discursively* produced. It is a discourse that both opens up and closes down numerous gaps of rejection and resistance in the culture at large. I am contending here that such a complex production must itself become the object of investigation if white theology is ever (theologically) to evaluate its own identity as white. A contextualized white theology would thus have to comprehend itself as not only "object" in the mirror of a Black Theology that objects *to* it. It would also have to comprehend itself as "speaking subject" in the very discourse it jointly produces with Black Theology. It would have to address both blackness *and* whiteness, simultaneously. Any white attention paid to Black Theology merely as "black" and thus "different"—no matter how appreciative—without correlative attention to the theological meaning of whiteness (as that which blackness differs from), already fails to take seriously the meaning of Black Theology's message.

Witvliet's proposal for a critical dialogical engagement between white and Black theologies is certainly a good beginning from the perspective of European ecumenical theology. Certainly within the context and content of his own (chosen) discipline as a European lecturer in ecumenism, Witvliet is allowing himself to be challenged and moved (Witvliet, 1987, 98). But he remains as a white largely circumscribed within an institutional order that has final control over the social geography and disciplinary boundaries of that encounter. Contrary to his own self-confession, "reading black texts" is not the same as "entering the world of blackness" (Witvliet, 1987, 265). Perhaps, as indicated above, there is currently no other possibility for a white European response to Black Theology.

But for white Americans, the "text" of blackness is a matter of everyday practices and performances. Whites are involved in blackness before and outside of ever reading a Black Theology text. We are involved before doing anything at all, but simply in virtue of our living as Americans. "America" as a set of social structures and cultural performances encoding both historical memory and political agency is already determined by the ongoing articulation of black and white values, experiences, presuppositions, biases, behaviors, styles, and habits before these latter are ever reified in academic texts as defined objects of thought and writing. The pervasive presence of this "already-ness" of race is the subject of chapter 4.

The Crisis in White Materiality: Salvation as Self-Aware Embodiment

But we are left with one last point to be made in this chapter before proceeding to the analysis in chapter 3 of the historical conflation of race and soteriology in America. And that is that discourse itself has an "exterior" that affects and effects its production of meaning. Whether one chooses, as Jacques Derrida and many poststructuralists do, to understand language as encompassing everything, as a "textuality" human beings can never get beyond, or chooses, instead, to speak of the "extra-linguistic ramifications of power at work shaping communicative acts," or the "non-discursive operations of power such as modes of economic production, state apparatuses, and bureaucratic institutions," as Paul Gilroy and Cornel West do, respectively—one must somehow account for the way materiality shows up in language (Gilroy, 1993, 57; West, 1989, 209; 1982, 49).

Here, I do not propose to develop a thoroughgoing philosophical accounting of such an "outside" of language, but rather to underscore some of the phenomenological effects that are at stake in relationship to the question of race. For the sake of ease of communication, I follow the course pursued by West and Gilroy rather than Derrida. I use terms like "pre-rational," "extra-discursive," "trans-linguistic," to refer to material and structural aspects of the human universe that exercise force upon and within the human body even when they are not precisely rendered in language. I do not mean that human beings can ever register the presence of, or refer to, these forces in other than linguistically coded ways. I do mean that the "text" human beings jointly produce and reproduce and inhabit socially admits of many levels and layers of meaning, not all of which are conscious, or coherent, or able to be articulated in the form of words. The body itself is a world of meanings, signifying the simultaneity of its history of desires and disciplines—its tensions and intentions, its actions, reactions, and chance eruptions of feeling and energy—in every moment of its psychosocial existence. And it can be *made* to signify more than one meaning, and on more than one level, at the same time.

My point here is that James Cone's written text does not reflect the full force of the racial history that his body refracts when he presents that text orally in one or another social or theological forum. And yet, I would contend that the meaning of Black Theology cannot be adequately treated, by white theology, apart from those other levels of meaning that are inevitably brought into play and played with so powerfully, in the oral performances of a Cone or many other black speakers. There are various extra-discursive levels of both Sameness and Difference that are so regularly "signified upon" and worked with in black expressive cultures, that they become a kind of spiritual formation of the black body.[11] Under such a pedagogical regime, the black body becomes a lived *habitus*[12] of racialized meanings. It is invested with a corporeal consciousness and a repertoire of gestural signs that have been sharply honed as part of the apparatus of survival for living in a social context where such a body is regularly endangered simply because of its surface colors and shapes. In short, black vernacular culture, historically, has provided black people with a performative space in which to experiment and work through racialized meanings, somatically and affectively, before their lives are on the line.[13] In consequence, a kind of "dramaturgical competence" is often developed that makes the black body itself a therapeutic site for countering racism. While this is covered in more detail in chapter 5, it does demand at least a preliminary statement here.

In one sense, in this writing, my subject is that form of white *subjectivity* at large in America that regularly takes blackness (in general, and "color" in particular) as its implicit *object*. I am approaching Black Theology as a "significant event" on the horizon of that subject – object structure that raises a question inside of the structure itself. But as a theological intervention, such a question remains also largely an academic event within a white-controlled institution. I accept that discursive venue, but at the same time I want to resituate the event, reinstantiate it, reimmerse it in its sociopolitical, cultural, economic, and discursive complexity and depth as a phenomenal moment, and ask how white theology can come to grips with such an event and let it signify in all of its phenomenological profundity. Doing so requires attention to both discursive and performative signs. I am suggesting that Black Theology, in its interpretation of Black Power as its own most immediate historical "datum," privileges recent urban uprising in particular and offers itself as both a theological reflection on *and an existential expression of* that complex event of social communication.

"Underneath" white theology's encounter with Black Theology is a profound sociocultural, historical, and archetypal depth that is lost if the encounter is allowed to remain only textual and academic. The urban upheaval of early 1990s Los Angeles is the signature event I invoke to signify that extra-discursive depth in its most massive and complex form of expression. But it is Cone's theology that represents the initial theological *articulation* of such a historical moment. What remains to be said and taken account of is the fact that Cone himself (as indeed, many other African American scholars and speakers) regularly reiterates something of the ferocity of that explosion in his own presentations.

Cone's actual oral presentation of his academic position regularly gives performative flesh to the historical blackness he writes about in his text. In a sense, my project here is to try to write that performative "extra" back into my own representation of Black Theology as a white exercise of self-criticism and a coming to partial self-consciousness. The performance in question is not to be dismissed as mere idiosyncracy. It is rather a performance that arguably (as we shall see in chapter 5) mobilizes a communal "structure of feeling" that can trace its genealogy all the way back to slavery. Its active reproductions and proliferations in the black community begin with a politics by ritual that transfigures the terrors of slavery into a resource for resistance. That original experience of terror is inverted in such historical counterterrors as the Haitian revolution of 1791–1804, repeated on a small scale in Nat Turner's "apocalyptic ride" through Virginia in 1831 and in other mini-slave revolts of the nineteenth century. The resulting political affectivity runs right through the vast prodigality of black aesthetic creativity in the spirituals and the blues and trickster tales and jazz; it shows up in the interdisciplinary depth of intellectual works like those produced by W. E. B. Du Bois and Zora Neal Hurston, Richard Wright, and Toni Morrison; it emerges in the rich radicality of activists like Frederick Douglas and Sojourner Truth, Marcus Garvey and Angela Davis, Martin Luther King and Fannie Lou Hammer; it reveals its gravity in the hard faces and harder rhetoric of the Nation of Islam both in the days of Malcolm X and more recently with Louis Farrakhan; and signals its unresolved contemporaneity in the explosions of Watts, Newark, Detroit, and other cities, in the 1960s and LA in the 1990s and the politico-cultural movements that grew out of that legacy of struggle such as Black Power and hiphop. Obviously, each of these historical developments is a complex phenomenon in its own right and cannot be reductively explained by some facile reference to a "structure of feeling" dating back to slavery. But I do follow Gilroy here in arguing that there is some critical element of expressive difference—identified as "black" in its own self-naming—that goes to the heart of various political questions of identity raised by the practices of racialization and racism in much of the Atlantic world of today (Gilroy, 1993, 37, 73, 129, 200).

In keeping with that agreement with Gilroy, I am insisting that Cone's formal christological claim in his academic texts puts the issue squarely where it needs to be, in the deep place of soteriology, and that the depths of that soteriological question cannot be fully perceived, much less understood, by whites, except in the form of an encounter with a black performative confrontation that puts them at issue not only historically and theoretically, but existentially and practically, in the here and now. The affective undercurrents of that encounter, for whites, arise out of the history of race codified *in the white body* as both fear and guilt—a fear and guilt the size and depth of the terror and anger conserved and configured in slave significations and black communal practices. I am thus suggesting that white corporeality is also rooted in a "structure of feeling" that is socially reproduced. But guilt, in human experience, is a slippery emotion. It lends itself to both projection and overdramatization—especially when coupled with power. The history of guilt in this country, the genealogy of affective structures that are characteristically "white," is a history of displacement and denial that will require deep subliminal work if it is to be uncovered, owned, and exorcized.

A white theology serious about the multiple social, political, economic, and cultural challenges of overcoming racism thus faces also an initiatory task. The struggle to respond honestly and adequately to Black Theology will cut all the way back to the deepest levels of personal identity, the most inchoate structures of bodily *habitus*. Such a theology is faced with the task of learning to relocate its tragic guilt, in perception and performance, back inside the white body, and mature, out of deep transformative work with that guilt, an interior capacity for compassion and creativity, rather than dissipating its energies in projection and anger or resignation and avoidance. But deep guilt is rarely faced and worked through on the basis of mere self-initiative alone. And here we return to the fact of black performance.

Cone's Black Theological "performances" signal much, much more than a mere facility with words and capacity for polemic. For white auditors, they demand the recognition that whiteness is *also always* a performance—and thus, not a natural fact, not an inevitability, not an unchangeable "given" (even if whiteness is usually a performance that is quite unaware of itself). Cone's oral assertiveness tacitly, but unavoidably, puts the body on the table as a *theological* issue, an issue of incarnation. His particular style "gives flesh" to the necessity to confront not only the cognitive contents of written blackness as specified in a formalized christology, but equally, the affective contents of performed blackness as specified in the forms of encounter. Those contents codify a profound and ongoing historical confrontation.

Part of Cone's gift is his capacity to embody that confrontation without apology wherever he presents, even in settings that are heavily "white" in their protocols of behavior. Such a performance already breaks open the structures of feeling of the encounter. The implicit demand of such a performance is for what I would call "white initiation." The task enjoined is not only mental and spiritual, but also somatic and psychic. The need is for a thoroughgoing de-sedimentation of the repressed content of whiteness. Of course, the full force of the confrontation—one that is, in fact, already present, but repressed, *within* white identity in its structure of fear—could only be made fully conscious in a reversal of social roles. Only a prolonged experience of "reverse minoritization," in which whites experienced occupying the place of the racial "other" without relief for an extended period, could accomplish a full initiation into the deep meaning of race. (And even then, such an experience could not begin to approximate what blacks face daily in this country as a constraint admitting no escape.) But Cone's public embodiment of some of the energies of racialized difference and contradiction at least opens up the possibility for whites of registering the "feeling-effects" of racialized terror and thus also the possibility of experiencing an affective resonance with black joy. Black expression, obviously, encodes not only struggle, but also ecstatic release.

Breakup of the structures of white fear and guilt cannot be accomplished in solitude. It requires self-confrontation before the memory of terror encoded in black expressive practices, and at least a beginning initiation into the possibility of creative healing such expressions permit. "Black" joy can create an open place inside the white body, whence emerges the possibility of self-knowledge. But here also lies the white demon—the dark root of guilt and fear in an insecurity whose most immediate expression is denial and dissimulation. What is required in place of denial is continuous self-confrontation, slow exorcism, and careful revision in a conscious resolve to live "race" differently. It is ultimately a matter of learning how to live creatively out of one's own diverse genealogy and experiment with one's sense of embodiment gracefully—*against* the dominating structures and conforming powers of white supremacy that have already conscripted one's body for their service. That hope is obviously utopic—but not to be abandoned simply because it represents the eschatological endpoint of white "salvation." Chapters 7 and 8 focus on this particular aspect of the predicament of whiteness.

Having said all of this at this point, however, there remains a danger that I will be heard saying that white salvation at the level of race depends upon black expression—that one more time, blacks are being set up to provide the "fat" of white living (this time, spiritually). I am not saying such, but I *am* arguing that there is a razor's edge that must be negotiated by white people for the foreseeable future in America—a difficult but rewarding journey between the twin dangers of self-sufficient "self-ignorance" on the one hand, and "other-dependent" appropriation and exploitation, on the other. The dialectic of dependence and independence that structures racialization and the conundrum of working out a new possibility of autonomy and exchange constitute the focus of the rest of the argument.

II

History, Consciousness, and Performance

The sacral edge of black cultural creativity, articulated by Black Theology, traceable in its origins in the New World back to slave innovations, cannot be adequately accounted for in merely discursive terms, but only by taking cognizance of a *difference of the body* in its negotiation of time and space in America. That body's disruptive difference from the hegemonic whiteness of Anglo-American Christianity can be evaluated theologically only in terms of a performative divinization: a *daemonized* Majesty, a syncopated Word, a "blues-bodied" Spirit. It institutes a "local knowledge" that is finally dramaturgically divined.

For whites, theological blackness became conceptually articulate[1] only with Cone and Albert Cleage and Jaquelyn Grant (and multiple others) in the latter part of the twentieth century. However, it has been practically clear for blacks from the beginning. It authorized a practice of black freedom as critical—as Nat Turner could testify—as anything envisioned by the Frankfort School today. Its text is the black body in motion, in community, often enough, incommunicado.[2] It is a body that has been "labored"—under all the terror and tension of W. E. B. Du Bois's double-consciousness—into both a tragedy of anguish and a triumph of overcoming that now stands on the horizon of America as a theological and political sign of the times. But such a recognition also points toward a necessary interrogation of the white body in its negotiation of time and space in history.

This latter task requires, as its first step, reconstruction of the genesis of modern racial supremacy as a *surrogate mode of soteriology* for the white Europeans who read phenotypic contrast and cultic difference into theological significance in their conquest of Others of Color in the various colonial theaters of New World contact. The section that follows here proceeds through these interrelated concerns by first addressing the *history* of white theological supremacy and the incorrigible black resistance it provoked, then sitting at the feet of Du Bois for a story of the *consciousness* such resistance evoked, and finally probing "under the skin" of black culture for a sketch of the *performative* competence that encounter with white terror has enjoined for a community determined to survive the onslaught.

3

Modern White Supremacy and Western Christian Soteriology

It is impossible for us to suppose these creatures to be men, because, allowing them to be men, a suspicion would follow that we ourselves are not Christians.

—Montesquieu (cited by Richard Drinnon, Facing West, 138–139).

On May 4, 1969, James Foreman interrupted the Sunday service at Riverside Church in New York City with a public proclamation of "The Black Manifesto," demanding $500,000,000 from white Christian churches and Jewish synagogues as reparations to black people for three centuries of exploitation and brutalization. The document he read represented the thinking of a new breed of militant black clergy who first began crystalizing their demands in discussions with each other in 1967. In one sense, the manifesto could be said to have represented the religious articulation of the unresolved rage and unrequited agony that had galvanized the southern "freedom movement" of the early 1960s and the northern unrest that attended its repression later in the decade. In style, if not in explicit content or conscious articulation, it owed as much to Malcolm X and the Black Panthers as to Martin Luther King and the Black Church. Black Theology became its legacy and James Cone its foremost theologian.

I begin this chapter with such a brief recap not only because it grounds the subject of this inquiry in history, but also because it portends the stakes. Foreman's intervention could be said to have represented a bold, if naive, gesture of prophetic symbolic action. It brought Civil Rights "sit-in" courage and Black Power "closed-fist" rhetoric into a new arena. In its challenge, for the first time, the white liberal

Christian church itself came under attack. The latter's own self-definition as the body of Christ was questioned. From the point of view of black Christians, the white church appeared rather as the battleground of the "principalities and powers."[1] White liberal fury, in response, was tangible. Here, for an awkward moment, the white-faced Jesus-mask slipped and revealed its cult-double: the ferocious tribal god of white supremacy.[2] Blond hair and blue eyes, it became apparent, could reference more than just cultural privilege. They could also signal spiritual possession.[3] Christ could hide an anti-Christ.

But if it is possible to discern profound spiritual combat in such a fraught liturgical scene, it is also possible to embrace the clarity of its human challenge. The Black Manifesto asserted a basic condition of American history. White well-being and black embattlement are agonizingly intertwined. Indisputably, whites owe blacks. In the heartland of liberal capitalism, the "political economy" of race relations shows itself as a debt-structure at virtually every level. And the ledger has yet to be cleared.

It is under the burden of this double conviction that this particular project goes forth and names its task. Spiritual battle and human struggle, contradictory symbol and conflicted community, theology in the idiom of race, race in the key of theology—the work to be done is complex and contested. The divinized subject of christology and the racialized subject of anthropology must be thought together. And there is no neutral place from which to analyze, no prospect of innocent engagement. Salvation, on the scene of race in contemporary America, remains captive to a history of oppression. It whispers its gift as Spirit; in the same breath it shouts obligation. And it does both in the present tense.

The Crisis of Race in the History of Modernity

At first glance, however, this insistence (on a theological treatment of race from the standpoint of soteriology) might seem either melodramatically excessive or theoretically confused. Soteriology deals with the "logic" of salvation, with one's status in relationship to the ultimate. Racialization, on the other hand, deals with something historically proximate, with a logic of relationships at the level of everyday life and practice. Consideration of the latter with respect to the former is not necessarily an immediately obvious theological move to make.

However, even a cursory glance at the postcolonial legacy of conquest and slavery reveals that the connection between salvation and racialization has never been an "outside" imposition of a theological category on a nontheological subject.

Context of the Crisis: History

Historically, soteriological questions have long figured in European determinations of the meaning of cultural difference in encounters with other peoples.[4] From a European colonizing perspective, for instance, America offered the promise of a kind of salvific purity—a chance to start over, leaving behind the contagion and compromises of a birthland wracked by religious wars and divine right pogroms. The Atlantic served as baptismal font, imposing its vast watery openness as rite of initiation, purging the old. The shores of Massachusetts or Virginia loomed on the horizon as the soils of Canaan, awaiting recognition as the Promised Land. And the native inhabitants emerging from the "wild" forests in queer garb and speaking strange tongues offered new "material" for the making of heaven on earth (Dussel, 1995, 35). Their role in the new regime of revelation would ultimately be decided by their placement in the economy of salvation.

More immediately, however, they posed a profound hermeneutic problem.[5] The evident difference represented by these indigenous peoples did not find ready placement in the grand narrative of interpretation providing the epistemological framework for European expansion and settlement. No ready-to-hand European cultural category rendered their sudden appearance on the Christian horizon intelligible. The lack of fit provoked taxonomic travail. It was a travail with a long history. While it is not part of the purpose of my writing here to attempt a broad survey or deep critique of the history of European responses to difference, it is important at least to set this conundrum of colonial classification in a temporal framework that provides perspective for the conjunction of race and salvation I am claiming for our contemporary moment.

The Crisis in Premodern Europe: Salvation and the Internal Other

In early modernity, racial discourse emerges as a species of theological evaluation as we shall see in what follows. European efforts to discern cross-cultural meanings in encounters with other peoples around the globe after 1492 issue in perceptions that increasingly connect immediate appearance with ultimate destiny. But this ready conflation of salvation and racialization does not begin *de novo* in that early hour of confrontation. It has its precursors in a whole set of categories within Hebrew, Christian, and Greek thought that inform older European attempts to deal with difference prior to the radically new experiences of the fifteenth and sixteenth centuries. In his *Tropics of Discourse*, historiographer Hayden White traces this earlier career of otherness primarily in relationship to ancient notions of "wildness." While his treatment is only one of many, it helpfully highlights the theological stakes.

In a chapter entitled, "The Forms of Wildness," White begins his sketch by noting that within the Hebrew writings, the category of "wildness" emerges as one of the primary vehicles for interpreting difference. It gathers its meaning in large part as the archetypal opposite of "blessedness." It is explained in those texts as a form of "species corruption" that results from accursedness and issues in an unnatural mixing of human and animal attributes (White, 160). One place where the operation of such a meaning becomes clearest is in connection with the figure of Ham. He becomes the eponymous symbol within the tradition around which an etiology of wildness is elaborated. In the Genesis accounts, Ham is (mis-read as being) cursed for revealing the nakedness of his (drunken) father, Noah. He is then carried forward in the tradition (or more accurately "retrojected" backward, since Genesis is constructed late in the traditioning process) as the ancestral bearer of such a curse, appearing in the genealogies as the progenitor of a breed of "wild men" combining Cain's rebelliousness and pre-Flood giantism (White, 161). Later descriptions of human anomaly in the Hebrew text often reference him as prototype. Indeed, the later tradition reveals an etymological conflation of words for "blackness," "Egypt" (as the place of bondage), "Canaan" (as the site of pagan idolatry), "accursedness" (coupled, ironically, with notions of fertility) and the name "Ham." In at least one Hebrew worldview, then, the "form and attributes of wildness" testified to a contamination that was understood to have been at once biogenetic and politicoreligious. It amounted to a well nigh "insuperable condition, once it had been fallen into" (White, 162).

The later Christian approach to difference, on the other hand, was, in theory, more charitable. In the official Church view, evident otherness was not locked up in insuperability (like it appeared to be in the Hebrew texts). The Incarnation marked a divine intervention in time that altered human anthropology. After the birth of Christ, every human being was deemed "salvageable in principle." "Whatever the state of physical degeneracy into which a [person] fell, the soul remained in a state of potential grace" (White, 163). Although according to Augustine, the Hamitic legacy meant the genesis of "monstrous races" as well as monstrous individuals, monstrosity itself did not exercise an absolute claim. In a theological worldview heavily influenced by Neoplatonic (and Aristotelian) categories, a subtle distinction could be made. In light of the differentiation of the essentialized soul from an accidental body, the difference between a seemingly monstrous appearance and a normal Christian (or even pagan) humanity was "one of degree rather than of kind, of physical appearance alone rather than of moral substance manifested in physical appearance" (White, 164).

But as White notes, there was a significant difference between official ecclesiastical theory and popular lay practice. While official doctrine insisted salvation was accessible through the church to any human soul, no matter the physical appearance, popular thought begged to differ. The commonsense thinking of lay Christians— especially among the still partly pagan peasantry—pushed beyond this idea of "mere" (and thus redeemable) "degradation" to the horrifying possibility of a human being endowed with an irredeemable animal soul. Unofficially, White says, the Middle Ages expressed, through popular legends of Wild Men and Wild Women, a "deep anxiety[6] . . . that one might regress to a condition in which the very *chance* of salvation might be lost" (White, 165). Not so much the means but the grounds of salvation might disappear. And while the retrieval of classical Greek culture in the Renaissance immensely complicated the scene, the question that "difference" posed to soteriology in the late Middle Ages remained very much present. In the late Middle Ages, "wild figures" (ancient satyrs, fauns, nymphs, *silenni*) were indeed refigured positively in popular culture as the "protectors and teachers of peasants" (over against official society). But that development notwithstanding, the specter of otherness continued to trouble social relations. At issue personally was the security of individual destiny; at stake politically was the very character of European culture itself.

The Crisis in Medieval Spain: Salvation and the Abrahamic Other

Nowhere perhaps was this question of European political character more heavily loaded with Christian spiritual anxiety than Spain of the

Reconquista. Muslim takeover of the Iberian peninsula (all except for a thin strip in the north) beginning in 711 CE launched a tri-religious experiment under Islamic hegemony that flourished for two centuries and then gradually eroded under onslaughts of conservative Berber Muslim challenge from the south and equally conservative Christian Castilian aggression from the north (Menocal, 267). As with virtually every recounting of history, the record is replete with complex conflicts and unanticipated cooperations, surprising turns of events and colorful characters of both good and ill repute (depending on who is telling the story). What is most important for our narrative here is the way political power and spiritual uncertainty come to be amalgamated in the struggle of Christianity with its most incorrigible religious "others."

Throughout the feudal period, European Christianity is a rather crude popular practice manipulated by rabid political interests under the ever-looming shadow of a vast and sophisticated Islamic empire encompassing all points south and east. Spain had become the terrain of most consistent intimate contact between the two, albeit on Muslim, not Christian terms. Relations with the Jewish tradition out of which Christianity had emerged as "bastard child" (of mixed Jewish orthodox and Greek pagan parentage) had been fraught since the first century. In Spain, the colloquy between the three Abrahamic faiths, sponsored by an urbane Umayyad rulership, greatly contributed to progressive thought in each of the traditions, but also contributed to deep antipathy between the two most in competition for power. For much of the 750-plus years of contact, Christianity labored in the position of cultural–religious inferior to Islam, more often taught than teaching, persuaded even to embrace Arabic as the medium of both thought and liturgy (Menocal, 69). The Mozarabic expression of Christian vision, succeeding an already distinctive Visigothic orientation, partook of a Spanish piety that was profoundly polymath and pluralized (Menocal, 143–144). The shared ethos, dominated by Islamic and Jewish intellectual genius, inevitably engendered a measure of Christian sycophancy, with all the self-loathing and secret *ressentiment* such conformity usually masks. On the other hand, even the shrewdest minds of European Christian persuasion located further north recognized the theological and cultural brilliance of al-Andalus (as Islamic Spain so named itself) and sent representatives to train and translate under its tutelage.

In the hard-scrabble politics of the seven-century-long Christian "reconquest" of Spain, pride of both place and faith dictated a ruthless Castilian suppression of the Muslim and Jewish "other" once the military outcome had been secured. The year 1492 marked a momentous shift. Not only did Columbus set sail west to circumvent Levantine Muslim power commercially, but Ferdinand and Isabella repudiated all pluralist cooperation religiously in requiring every Muslim and Jew remaining in their newly proclaimed kingdom to either convert or leave. The *reconquista* had been waged under the twin banners of a militarized Marian devotion to Our Lady of Guadalupe (destined in short order to become the chosen champion of the *mestizo* offspring of Spanish rapes of Azteca and Tolteca women in all subsequent Mexican struggles to resist colonial oppression) and a refiguring of St. James as the "warrior supreme" (Dussel, 1980, 45–46).

The supremacy posited of the "horseback saint" (as James was often depicted) was thoroughly subjectivized in the male Spanish psyche in the reconquest's triumph (Pike, 434). Christianity had seemingly proven itself ascendant over its only serious Western-World competitor, Islam—and that in spite of Islam's clear intellectual preeminence. The peculiar experience of a long-endured cultural inferiority "trumped" now by a hard-won "spiritual" superiority quickly amalgamated into a ruthless presumption of supremacy that profoundly influenced Spanish colonial ventures (Dussel, 1995, 13). Columbus's journals reveal his own private ruminations on his voyages as more pilgrimage than simple commercial venture (Wessels, 57–59). He took his birth-name of "Christ-bearer" ("Christopher") eminently seriously. Part of his self-conceived role was extracting enough ore from the lands of discovery to fund a resumed crusade against the continuing "infidel" occupation of the Holy Land. Encounter with New World *indios* was mediated by the continuing specter of the "Moorish" other, so recently subdued in Spain itself, so threateningly on the rise in the east (Constantinople had fallen in 1453 CE leaving the Balkans utterly exposed to the Muslim expansion that would become the Ottoman Empire).

In the late Middle Ages, Christian questions about the certainty of salvation were thus animated not only by anomalous experience inside the community (the challenge of wildness detailed by White), but by the ongoing engagement with pervasive Islamic power at the insecure borders of the community. Christianity in Columbian confession exhibited a *pathos* of salvific confidence that belied in bellicosity what it asserted in theology. This was a supremacy still struggling to believe its own soliloquy.

The Crisis in the New World: Salvation and the "Savage" Other

In New World ventures, the European struggle to classify those it ruthlessly conquered and relentlessly colonized was underwritten by these unresolved spiritual anxieties and political contradictions on the home front.[7] The process was heavily weighted with metaphysical concerns and had grave existential consequences. Projections about the capacity of the other "to be saved" became a crucial qualifier in what quickly emerged as a kind of conundrum of the colonizer, a dilemma of the duty to evangelize and civilize.[8] On the one hand, if comprehended as "save-able," then the "wild savages" of these new lands were *de facto* equal to the colonizers as potential spiritual subjects of the Christian message and political subjects of the king. But if potentially equal in the economy of salvation, then how could such souls legitimately be exploited as slave-labor, or destroyed as heathen? On the other hand, if judged as somehow beyond salvation, where did these "Indians" fit in the great Christian cosmogram, how could sense be made of their capacity both to aid and to resist the colonizers, and what was their status in relationship to God's power to transform? In Europe's earliest attempts to decipher the significance of its others in the conquest of new lands, soteriology became, in many cases, *the* decisive category of classification, *the* open question around which various trading, colonizing, and evangelizing initiatives organized their competing discourses of legitimation.[9] Evident difference found its adjudicatory point in the discourse of salvation.

This was even more crucially the case with the earliest European initiatives in Africa, and later when Africans were transported to the Americas as slaves. On the question of the soteriological status of blacks hung much of consequence in their socioeconomic and politico-cultural placement within colonial affairs. Indeed, it is here, in its uneasy "discourse on blackness," that colonial soteriology achieved its most intransigent articulation with race. Black skin posed the question of salvation in its starkest form. And nowhere was it posed more poignantly than in South Africa and (North) America.[10]

In the particular (and quite unique) cases of the indigenous inhabitants of South Africa and the imported slaves of (North) America, pigmentation was made to conflate with Calvinist notions of predestination. As social scientist Roger Bastide has noted, in these two situations, black refusal of the Christian message outright—or what was just as bad, unapologetic mixing of Christian categories with traditional African practices—was taken by whites as a visible indication of a diabolical affiliation.[11] Just as white "economic success was proof of divine grace," so black recalcitrance became "the sign of their rejection" (Bastide, 281). In a Calvinist symbology that tended to festishize the signs of election, blackness began to signal a predilection for damnation. Its opacity could be read as the "transparent evidence" of a resistance to God so thoroughgoing it had seemingly reproduced its meaning on the surface of the body (Bastide, 272). Dark skin was made to prefigure a destiny of perdition.

At the same time, this patently soteriological "discourse on difference" had profound social effect in both America and South Africa. It signified the ultimacy it read into blackness with visceral ferocity and clear policy. For many colonial and postcolonial whites, formed in a culture conditioned by the ecclesiological implications of Calvinist notions of election and perseverance, the import of such a skin-sign was unequivocal. In his own day, Calvin had expressed explicit concern (in the *Institutes*) for the kind of temptations that could threaten the elect merely because they opted to live the country life where (some) contact with "savages" was unavoidable. How much greater the worry where the contact was more constant and the sign of savagery (seemingly) more overt? In both America and South Africa, the epidermal evidence "spoke" with unambiguous force. It enjoined social and psychological segregation. The elect sought to protect themselves against the possibility of perversion and pollution at every turn. At the same time, given Calvin's limitation of the precept of salvation to those who had some alliance or affinity with Christians, *white culture*, in effect, became identified with both the defense of the faith and the demographics of health. It established the borderline of both spiritual and material certainty. Only within whiteness was one's existence secure. Bastide characterizes the consequences succinctly:

> In this way, dark skin came to symbolize, both in Africa and in America, the voluntary and stubborn abandonment of a race in sin. Contact with this race endangered the white person's soul and the whiteness of his [or her] spirit. The symbolism of color thus took on one of the most complicated and subtle forms, in both Protestantism and Catholicism, through the various steps through which darkness of color became associated with evil itself. (Bastide, 281)

In relationship to American racialization in particular, we could summarize as follows. In the dominant (white) Euro-American cultural discourse of colonial (and early postcolonial) America, black skin marked a profoundly theological boundary. It figured a *soteriological* threshold beyond which Christian destiny became dangerously uncertain. It threatened potential contagion for a salvation figured as "white." At the same time, Puritan (Calvinist) constructions of election helped mark both blackness and whiteness as *socially* significant in a specific manner. They underwrote the particular way in which slavery was instituted in the United States and helped leverage the various disciplines and structures of exclusion that white people have practiced upon black people, individually and collectively, ever since.[12] In a double operation, blackness and whiteness together articulated an everyday boundary for an ultimate concern. At the same time (in a reverse effect), the latter (ultimate concern) loaded the heavy weight of eternal significance into each of the former.

Which is to say, in America—in a historically unique manner—racialization organized social differentiation by means of soteriological signification.[13] It is part of the argumentation of this work that the effects of that articulation remain visible and virulent up to this very moment—even if the discourse itself on race has publicly shifted into a secular key.[14] But "adjudication of difference" was not the only soteriological intervention in the colonial struggle to classify and control in America. Questions of salvation also irrupted inside the adjudicating subject.

The Crisis Among the Colonizers: Salvation and European Sameness

In his book, *Prophesy Deliverance!: An Afro-American Revolutionary Christianity*, Cornel West makes the point that the dilemma faced by African Americans in the colonial period of America was actually one of a "triple crisis of self-recognition" (West, 1982, 31). Not only did African Americans have to negotiate continually the painful "double-consciousness" later articulated by W. E. B. Du Bois (that shall occupy our attention in chapter 4) but they also had to do so in the middle of a "broader dialectic" in which Americans struggled with "being American yet feeling European, of being provincial but yearning for British cosmopolitanism, of being at once incompletely civilized and materially prosperous" (West, 1982, 31). Obviously, this broader American dialectic was primarily a "white" dilemma. It amounted to an uncertainty or equivocation of self-identity troubling those who had left Europe without leaving behind all of their European-ness.

Here again, soteriological concerns created and maintained a border of difference. In this case, however, the difference appeared inside of Christianity itself. To the degree Europeans came to America in search of *Protestant* religious freedom, their soteriological *ethos*, if not their explicit schemas of salvation, effected a tremendous ambivalence. In the name of seeking to secure their place inside the domain of salvation, they *de facto* placed themselves far outside of the privileged social collectivity. They became themselves socially and geographically "other" to accomplish the spiritual goal of their (European) Christian convictions. In this intra-European domain of identity (between Anglo-American colonists and Anglo-British natives), the question of salvation did not serve to reinforce a narrated difference between people in a situation of geographical proximity (such as that between "Christians" and "savages" in the colonies, or later on, between white masters and black slaves on the same plantation). Rather, it helped to motivate a geographic (and spiritual and social) removal of some Europeans from other Europeans within a narrated sameness (i.e., the sameness of having come from Christian Europe).

In the process, soteriology itself was rendered equivocal. The Puritan vision projected a hoped-for spiritual fulfillment that had transported itself, for the sake of its own realization, outside of Christendom proper. The "way" of salvation—its earthly route or space of practice—now lay within the "wilderness" of the New World. But at the same time, its subject yet remained avowedly inside European *culture*, anxiously invoking an identity ambivalently fixed, in its point of reference, beyond itself, across the Atlantic.

Said another way, the predominantly Anglo Protestant colonies of America represent a peculiar spatial instance of the split of European society and identity that marks the advent of modernity.[15] In a certain sense, the Reformation itself could be, and often is, claimed as the evident beginning of the modern period. In its antagonisms appear the first signs of a complex historical fracturing whose causal "origins" remain in dispute, but whose effects are patent. The post-Reformation wars of religion issue eventually (in the 1648 Peace of Westphalia) in an exhausted religious *detente* and a strident demand to (re)found European society on a rational contractual basis outside the authority of a strictly religious vision. And while it is possible to argue for an earlier and broader beginning of the antinomies of modernity— namely, the fracturing of narrative time and social space occasioned by the "discovery" of other places and peoples with other gods and goods[16]—the Reformation remains decisive for Europe's own self-consciousness. In Reformation confrontations, the taken-for-granted holism of European social identity comes irrepressibly into conflict—with itself. The sacred correspondence of individual destiny and common good in the social synthesis imagined for itself by Christian Europe in the Middle Ages cannot be maintained. It is conclusively shattered as the operative ideology in the Protestant–Catholic struggle for politico-religious hegemony of the late sixteenth and early seventeenth centuries. In its place, self-conscious religious confessionalism emerges as the first language of European breakup.

But at the same time, it is also patent that the anxiety-ridden[17] drive for salvific security characteristic of the late Middle Ages does not just dissipate in that seventeenth-century dissolution. Indeed, as the known size of the universe expanded to infinity, and society lost its vision of a common destiny, anxiety over security found itself concentrated evermore intensely within the individual subject. While it is beyond the scope of this study to develop a full-blown thesis about the career of what I would call "soteriological anxiety" in modernity, it is possible at least to intimate an *ethos*. It is arguable that the late medieval "drive for certainty" enjoyed more than just an entropic half-life in what followed. Its traces can be found perhaps most suggestively in the near absolutism with which subsequent European struggles over difference (between peoples, nations, religions, classes, etc.) were usually conducted. Both revolution and romanticism recapitulated the need for an "absolute of identity" upon which to stake everything.[18] Colonialism's "drive to civilize" at least refracted, if not reflected, the earlier Christian obsession with likeness,[19] while capitalism's "drive to globalize" formalized the Calvinist soteriology of success.[20] The birth of the novel genre in the eighteenth century exhibited a quasi-soteriological quest-structure, refigured now for an individual subject, searching for meaning in a social world too fragmented to provide the coherence sought for.[21] Indeed, Enlightenment rationality itself, insofar as it presupposed universal articulation, claimed objective vision, and set about the task of a totalized classification, could be said to have drawn into its own subject of thought many of the compulsions to inquisition, certainty, and catholicity of the earlier religious synthesis.

The Crisis in the Colonies: Salvation and American Distinctiveness

Within this ferment, "America" appears as one particularly unique case in point. What will eventually become the "united states" begins to emerge, during and after the European wars of religion, as part of a transatlantic colonial orbit that can only be described, in relationship to its colonized others, as an ambit of cultural sameness (in spite of the differing makeup, cultural background, religious beliefs, etc. of its various Euro-immigrant populations). The 13 colonies work out the intensity of their Puritan-dominated mix of identities in a complex negotiation with both their environment and their forebears. Not only did the settlers' vision of an "absolute" form of salvation translate into nearly absolute genocide for the indigenous population but it also helped galvanize a nearly absolute pursuit of individualized well-being. In its drive for independence, America could be said to have realized in reality the new European demand for a political arbitration of religious commitment by reason. It constituted itself as a politico-geographic community-of-identity inside of an ambivalent cultural similarity that nonetheless was committed to hosting a plurality of soteriological aspirations. In its "Great Awakenings," the emerging social order could be said to have become the "national" site of an increasingly localized quest for fulfillment. Those developments both fostered and furthered the narrowing of the scope of salvation to the individual. American frontier ideals of independence and self-sufficiency merely solemnized the shrinkage as "virtue."

Only in its Civil War was America forced to complete its eminently liberal revolution with a concern for the social. And even then, "justice for the slaves" elbowed its way into the trinity of sacred pursuits (of life, liberty, and happiness) as only a *late* justification for the blood-letting of apocalyptic proportions. The fighting had actually commenced in the name of stopping the breakup of shared white interests (and not as a war over abolition of slavery).

For all of this nation's prodigal experience, however, it remains unclear today whether anything of America's original intensity, or its long-standing ambivalence, have lessened. The country may indeed have kept its religious confessions separate from its levers of control. But as modernity progresses, the boundaries of American self-certainty remain visibly troubled by quasi-religious "absolutes." Whether in the form of nineteenth-century Americanism, or twentieth-century Klan terrorism, whether in the 1846 takeover of Northern Mexico, or the 1964 (attempted) "take-down" of North Vietnam, as far afield as the "republican" (!) imperialism exercised toward the Philippines in 1898 or the "unipolarist" imperialism seeking to exorcize Iraq in 2003, as close to home as the forcible reeducation of natives earlier this century or the forceful misapprehension of "illegals" of late—the definitions of where "America" stops and the "other" begins remain tendentious and vicious. These definitions have not become less soteriologically fraught for having been secured by more secular forms of identification than those of Puritan precursors.

But neither is America alone in the evident terror of its uncertainty. Indeed, we would have to say, the demons of difference have continued to provoke the holy wrath of anxiety across the entire landscape of modernity. Yet the anxiety of the age does exhibit its own unique specificity. In modernity at large, the struggle for identity continues apace as a battle for one form or another of historical "absolution" or quasi-utopian wholeness. But now the categories it invokes—and the coercion those categories serve to legitimize—are scientifically secured.[22] Here we can only briefly sketch what others have so thoroughly exhibited before turning our attention to specifically "black" ways of dealing with the bedevilment.

The Crisis of Race in the Theory of Modernity

As already indicated, in the aftermath of the intra-European Peace of Westphalia resolving the wars of religion into an uneasy *cuius regio, eius religio*, soteriology ceases to be able to image itself in society. It can neither count on a totalized social approximation of its imagined wholeness nor command an unquestioned trajectory for the individual believer.[23] Salvation itself becomes the object of warring camps. Its discourse emerges as the first "ideological" casualty of a rampant process of fission and breakup.

By the beginning of the twentieth century, in the likes of a Max Weber, Western social theory will begin to project its object of study as unimaginable in any totalized, holistic sense.[24] By the last decade of the century, social theory in general—including various revised Marxisms—will concur.[25] The subject of study will be comprehended—under ideal-typical constructs—as dispersed across the various autonomous spheres of science, morality, art, and religion (and politics and economics and erotics, etc.), each of which struggles against the others in an irresolvable "war of the gods." Indeed, in a sense, in the modern period, salvation itself becomes simply one sphere of concern among many others for any modernized individual or community. (We could perhaps even speak, after Nietzsche, of a "will-to-salvation" operating in some circumstances as the religious form of the will-to-power.)

The hope and imagination of an absolute wholeness is (by definition) accorded a merely relative and assailed value.

But recognizing such does not also imply the disappearance of soteriology's premodern powers either to create or adjudicate social differences that are freighted with the weight of religious ultimacy and urgency. Neither pre-1945 Germany nor post-1969 Northern Ireland, not South Africa of late nor Palestine today, permits such a thought. Secular rationalization and religious salvation remain profoundly conflated in both politics and practice.

The Crisis in Theology: Salvation as Racial Identification

If we specify *salvation* as a given society's "intimations of ultimacy" glimpsed in the way it images human wholeness for itself (as those images show up in its symbolic codes, its community structures, and its ritual expressions), and *soteriology* as both society's discourse on, and its drive for, the absolute (i.e., as both its discursive *logos* and its "logic of practice"), we can indeed trace soteriological currents in modernity. In doing so, I am obviously begging the question of what constitutes ultimacy. My only point at this juncture is that, functionally, "absolutes" continue to leverage decisions about life and death in modernity. This continues to be the case whether they are understood in terms of a liberal Kantian "categorical imperative" whose demands are consciously adjudicated inside the individual subject, a more socially determined "bad faith" that operates (somewhat) below the surface of conscious decision-making in Marxist, Freudian, or Nietzschean constructions of capitalism, sexuality, or culture, or a more discursively constituted "effect of power" celebrated, negotiated, or otherwise dissipated, in today's fragmented, postmodern subjectivity. As such, I focus in what follows on articulating soteriology in *both* of its aspects: as a practical *logic* (or a drive toward an absolutely held hope of wholeness) that gives rise to (and is animated by) a *discourse* on that drive.

Which is to say, I want to understand soteriology very broadly here, as any (political) logic that discursively legitimizes a choice to risk the "human absolute"—the suffering (or causing) of death—for the sake of the preservation or accomplishment of a pure or whole identity. In this formulation, I am simply "thinking with" the master/slave encounter theorized in Hegel's *Phenomenology of Spirit*. If Hegel's overall dialectic can be characterized—as it has been by Cornel West—as "the form of a Christian Christology gone mad," I want to push the characterization one step further, and understand the "struggle-to-death" that gives rise to the master/slave dynamic as the form of a Christian soteriology gone modern (West, 1982, 33). In the same Hegelian move in which christology is dialectically "spiritized" in and through history, soteriology could be said to have been anthropologized. It realizes the older soteriological binary of salvation and judgment, wholeness and rupture, life and death, that structured the relationship between a heavenly Lord and an earthly servant, in a new dialecticized structure of consciousness. It is now in the struggle of consciousness to become self-consciousness by forcing recognition of itself by its other that we find an absolute power at work. It realizes its epochal moment in the choice either to risk death for the sake of freedom or to submit to bondage for the sake of security that is allegorized in the master/slave conflict—a choice that, for Hegel, is constitutive of modernity in both of its outcomes.

Like so much of Hegel's dialectic, this combat of consciousness is philosophically described in a manner that sublates, and thus preserves, its earlier religious formulations, and leaves (suggestively) indeterminate its precise psychological and sociological processes. Not surprisingly, as we shall see below, this indeterminacy has been creatively appropriated by a number of theorists in this century (e.g., Paul Gilroy, Frantz Fanon) to come up with counter-phenomenologies of the role of slavery *vis-à-vis* modernity. Here I am merely suggesting that to the degree this master/slave dialectic plots the social structure of an ultimate risk-taking in the drive of Absolute Spirit to its final reconciliation, it could be said to represent the subsumption, in the modern moment, of the past history of soteriology.

In what follows, I am positing that a soterio*logic* is at work (whether consciously or less-than-consciously) any time concrete forms of *identification* are being produced with respect to particular political options that are willing to risk the suffering or causing of death for the sake of achieving some form of projected wholeness. (This would hold true, even with those religio-philosophical systems like Buddhism or Vedanta that disavow "death" as "ultimate.") Obviously, the emphasis here is on the formal logic rather than the particular symbolic contents of any given soteriology. In this view, a politics thus becomes soteriological at the point where the identification absolutizes itself in its willingness to face or somehow justify death in the name of securing the boundaries and fulfillment of that identity.

It is necessary to add, however, that "absolute," in this formulation, does not mean "total." Indeed, differentiating between a process of absolute identification and one that claims to be total is perhaps one way of characterizing the modern predicament. We live in a time and space in history when some of our commitments may seem to demand everything *from* us without being able *to* promise everything to us. Obligations can arise (or are constructed) that face us with the possibility of a wholly obliterating, "absolute" death without offering us more than a fragmentary, perhaps transient, redemption or fulfillment. That is to say, modern obligations often seem to ask for an absolute existential commitment to what appears to be only a relative (not totalized) historical project. They leave us split between an absolute destiny and a partial identity.

In speaking this way of soteriology, it may seem that I am actually describing something closer to ideology. The resemblance is not adventitious. Indeed, in its rationalization, one way or another, of the fact of death, I would contend that all ideology is in some measure soteriological. And in helping to constitute concrete practices in history, all soteriology, even in its ancient or medieval forms, is ideological. But here, I cannot afford to get bogged down in attempting to clarify an exact modern demarcation between these two.[26] My focus is rather about a different kind of modern demarcation—one that concentrates on the process of difference-making and difference-marking itself.

Suffice it at this point to say that I do think deep questions about redemption—in or out of history—remain powerful and productive in the conflicts of modernity. Deep anxieties and desires about the prospects of realizing some kind of wholeness *that is also somehow uniquely and identifiably "one's own"* continue to operate. "Salvation" has not lost its potency for having lost its shape or even its subject. However we choose to define it, the drive for identity in modern cultures shows itself more and more intransigently as a flashpoint for social violence. It is ringed round with the fears and fantasies of ultimacy. It grapples with a form of "death" it can neither comprehend nor avoid.

Thus I would argue that, in spite of their secular forms, modern attempts at self-differentiation remain vexed by a soteriological desperation they can neither easily express nor finally believe. They remain "absolutely" inarticulate, but not unmoved. Not unpredictably, they show themselves vulnerable to a tendency to fetishize various this-worldly forms of imagined wholeness that appear, in modernity, as the political surrogates of a formerly religious redemption. Ethnic identities, I would argue, emerge in this context as one of the more (in)credible reconfigurations of the larger eschatological self lost in the modern disenchantment of the world. They at least allow for a kind of historical transcendence of the truncated spatiality and abbreviated time of modern experiences of subjective existence. Where such identities are invoked, purity becomes their essential criteria and race their most substantial form. Within the pathos of this modern predicament of subjectivity, then, the discourses of racialization could be said to give definitive articulation to a latent dream of absolute identification. In their various forms of practice, they constitute themselves as a politics of salvation.

The Crisis in Theory: Salvation as Scientific Rationalization

Said more succinctly, in the centuries-long transition from premodernity to modernity, the historical forms of a European quest for salvation have lost their *mythos* and their *kosmos* (their "cosmology"), while their *logos* (their "logic" or drive toward "necessity") and their *pathos* (their "structures of feeling" or "memory of suffering") remain vibrant and virulent. The effects of this simultaneous loss and continuation remain patent and potent in various modern processes of racialization and racialized conflict. The quintessential question of a premodern Christian worldview, addressed to human "otherness" in the situation of colonial contact, is radically transformed, *but not entirely soteriologically diluted*, in the modern encounter with difference. The question "Are you saveable within the spiritual economy of a redeemed Christian humanity?" becomes in Enlightenment form, "Are you orderable within the scientific taxonomy of a civilized European humanity?"

The quest for conversion, here, is replaced, not simply by a drive to civilize, but also subtly, by an impulse to classify (as we shall see below). There is an inversion of orientation. Assuring one's name is written in heaven is transformed into writing the other's name on earth. Objective truth goes bail for the truth of one's life. In a certain sense, the absolute difference between earthly struggle and heavenly wholeness is brought within the orbit of this world as a difference between subject and object.[27] But like an unstoppable run in a stocking, this "difference" unravels all the way back inside of the subject, as a form of "self-difference." Kant's distinction between phenomenal and noumenal realities, between the self as part of the system of natural necessity and the self as an inscrutable origination of freedom, marks the this-worldly apparition of what was formerly an otherworldly differentiation.[28] And "formal critique," here, arguably becomes the functional Enlightenment equivalent of medieval architectonics of salvation. It constitutes the new means of securing the foundations of existence.

Genealogically, I am asserting that this new theoretical incarnation of the "absolutes of differentiation" can trace its beginnings back to the late Middle Ages. The conditions for the possibility of such come into being with the shattering of the soteriological correlation between social identity and heavenly destiny that began with exploration and culminated in reformation. The early modern "discovery" of scientific objectivity, produced in its compulsion to classify, finds its motive force and affective necessity here. It transfigures an *angst* into a *kundst* of peculiarly modern provenance. Scientific investigation breaks apart the participative relationship to objects found in liturgy (the sacramental sensibility of the Middle Ages) and realizes itself *de facto* as a surrogate of worship.[29] And the classification that is racialization becomes one of its most (sociologically) fraught—as well as (psychologically) formative—operations.

Many social theorists and cultural critics of the last 20 years have detailed the historical emergence of scientific notions of race in post-Enlightenment Europe and its accompanying institutions and cultures of white supremacy. Yet others have demonstrated that the coincident appearance in Western history of scientific rationality and colonial racism is no mere coinci*dence*, but rather profoundly congenital and correlational. What is portentous in this pattern is the degree to which the conflation between race and reason becomes definitive of white subjectivity and the degree to which that subjectivity becomes decisive as a surrogate form of soteriology. Cornel West's analysis offers a useful delineation of the process.

The rise of white supremacy as an "object of [scientific] discourse in the West" traces its genealogy to the very "structure of modern discourse *at its inception*," according to West (West, 1982, 47). White supremacy is not simply determined by, but is constitutive of, the *episteme* of modernity. It finds its epistemological home in a set of mutually ramifying discourses that are entirely incapable of even "fielding" the idea of black equality. Scientific revolution, Cartesian philosophical innovation, and Renaissance revitalization of Greek ocular metaphors and classical aesthetical ideals together articulate an understanding of the world that has modern racism as one of its irresistible effects (West, 1982, 53). Supremacy is so deeply imbedded in this discursive formation as to be (seemingly) inseparable from it. How that might be so is a function of the peculiar interweaving of these "master" discourses.

Descartes, for West, marks not only the well-remarked "turn to the subject" in modern history, but also a fateful conjoining of the scientific intention to predict phenomena and the philosophical desire to represent such (West, 1982, 82). His famous method partakes of both. The philosophical concern for representation itself, however, is profoundly informed by fifteenth- and sixteenth-century retrievals of classical norms of beauty and proportion that are increasingly being invoked in service of an emerging (white, male) "gaze" that seeks to order all it surveys. In a reciprocal manner, the activity of this surveilling eye is focused by a "predilection to observation" that gives rise to the science of natural history and embodies an intention to order all physical bodies and visible qualities of things into the various taxonomies of the "one [great] chain of universal being" (popularized especially by Linnaeus) (West, 1982, 54–56). From these developments, the more formalized theories and practices of racialization come into being in the "comparative disciplines" of *physiognomy* (facial comparisons), *phrenology* (skull reading), and *prognathism* (facial angle as an index of character and soul). Science influences aesthetics, which informs philosophy, which represents science, which projects . . . race. Supremacy, in the mix, is a Gordian knot of premises— rooting its reading of race in *ontology* and *biology* and imagining itself as the epitome of the condition toward which black being (at best!) might aspire after "civilizing" uplift and integrationist assimilation. The view with Europe on top and others "under" repeats itself in all the "best thinking" of the age from Montesquieu, Hobbes, and Locke, through Hume, Kant, and Hegel, to Voltaire, Mills, and Jefferson.

Further intensifying this already formidable conjuncture is the fact that modern white supremacy was articulated in the service of global capitalism (West, 1982, 65). Its constitutive discourses were not only inseparable from, but were materially co-constituted *in*, the various imperialistic practices of a burgeoning European productive formation. Racialized ranking of others was and is an integral part of Western capital's ability to commandeer labor resources and raw materials around the globe. Supremacy warrants the rearrangement of indigenous markets and the extraction of indigenous goods as a necessity of "development" that finds its chief protagonists and beneficiaries in Western economic interests. The whiteness that presides over the rearrangement is both rationale and result. It is at once the favored subject of political-economy and the normative subject of racial theory.

This compound ramification of white racial privilege edges close to a covert meaning of soteriology. As the incorrigible effect of a complex discourse and coercive politics, modern white supremacy offers its subjects a fulfillment that is at once invisible and substantial. As an imagined ontological "presidency" at the apex of the great chain of being (or the great evolutionary schema), the payoff is psychic and subtle; as a concentration of (global) resources in a (local) suburb, the benefit is concrete and sensual. In this historically unique modality of subjectivity, the objectivity fetishism of science meets its material embodiment in the capitalist fetishizing of objects. Each of these pursues an aim that is virtually infinite—in the one case, a drive to classify that today seeks a unified field theory of . . . everything (!); in the other, a drive to accumulate without limit (!). White skin serves as shorthand for "right of recognition" in the schemas of the former and "right of entitlement" to the wealth of the latter. But what remains unaccounted for is the fetishism itself. Why the pressure to hallow social production with an absolute signification (Marx's mystified commodities)? Whence the need to metaphysicalize a taxonomy (Hegel's absolute spirit)? I have no pretension to explaining such here, but again only wish to keep the phenomena open to the question of the drive for wholeness.

The Crisis in Practicality: Salvation as Slave Jubilation

But "scientific" white supremacy is not the only modality of soteriology operating in modernity. Although he does not use the exact term "salvation," the work of British culture critic Paul Gilroy traces the effects of what could be called a surreptitious "second-life" of *religious* soteriology in modernity's nascent formulations of race. In *Black Atlantic: Modernity and Double Consciousness*, Gilroy argues that modernity itself is constituted in and by a split best comprehended through an examination of modern "race slavery." Unless modernity is understood from its underside, it is misunderstood. It is only in the survival struggles of populations forced to supply the wherewithal enjoyed by the "enlightened classes" that a subliminal meaning of modernity comes clear. Here modernity is revealed as terror and its successes as suffering. The achievements so roundly celebrated in Europe's own self-awareness are leveraged by relentless appropriation and ruthless suppression. Modern subjectivity, Gilroy argues, is constituted in a racial subtext of profound violence. But only the ritual ruminations of the enslaved return that terror to expression. This "dark" eloquence he calls the "slave sublime." I call it augury. But whatever the interpretation, the discourse of race as a practice of destiny in modernity finds its ultimate—and paradoxical—touchstone here.

What Gilroy proposes, in a critique of contemporary social theorists like Jurgen Habermas and Marshall Berman, is a reconstruction of the "primal history of modernity" from the slaves' point of view, based on Hegel's notion that slavery is modernity's very premise (Gilroy, 1993, 55). Both slave consciousness and slavery's coercions demand careful attention. From this perspective, many of the key issues of the modernity debates—history as progress, the idea of universality, the fixity of meaning, the coherence of the subject, the foundational ethnocentrism of bourgeois humanism—are found to be radically queried in the vernacular critique of racism articulated in slave *practices*.

Of particular interest in this slave vernacular, for Gilroy, are the primary *categories* utilized. In developing a practical elaboration of the critical distance between themselves as human beings and their condition as slaves, slaves did not immediately employ rational argumentation. Humanist ideals were not their first weapons of choice. Rather, what shows as primary are the *spiritual* ideals of an eschatological apocalypse—the radically revolutionary tenets of the Jubilee (Gilroy, 1993, 56). The slaves turned to the biblical categories that had been forced upon them and reoriented them to their own purposes.

But it is not just the categorical form that is of import. This self-conscious redeployment of premodern images and symbols gains an extra quantum of power from their mobilization within the brutal conditions of modern slavery. The symbols serve not merely as a utopian prophylactic against racialized terror. They actually conserve and nurture the remembrance of terror as itself a form of "slave sublime" that gives rise to an alternative consciousness of freedom (Gilroy, 1993, 37, 55–56, 77, 217–218).[30] Critical here, are the prediscursive "structures of feeling" preserved (from the experience of slavery) in these various Christian symbols and church practices that are revisited by ritual means in the ex-slave communities. These become—especially in the shared musical practices of black expressive cultures all around the Atlantic—a primary resource for the ongoing negotiation with modernity and its covert rationale of violence that slaves and their descendants have always had to face. In Gilroy's thinking, "terror" is thus present in modernity not only as the forgotten foreground of the visible Enlightenment, but indeed also as the "volatile core" of modernity's "insubordinate racial countercultures" (Gilroy, 1993, 73, 129, 200). The intensity of the experience of radical fear remains unspeakable, but is not thereby "unexpressed."

At the level of theory, these ritual rehearsals of intense memory signal both a critical limitation and an unspoken collusion. For Gilroy, slave practices contributed to the "formation of a vernacular variety of unhappy consciousness" (Gilroy, 1993, 206). From this point of view, the rootage of this consciousness in a "primal encounter with fear and vulnerability" suggests that criticism of modernity "cannot be satisfactorily completed from within its own philosophical and political norms" (Gilroy, 1993, 206, 56). Genuine critique demands rather that we rethink the whole apparatus of modernity precisely in relation to the conditions of slavery and the counter-politics it gave rise to. Patent, in such a politics, is an expressive style of communal antiphony that practically refuses the Enlightenment separation between art and life, ethics and politics. Patent also, however, is clear evidence of a thoroughgoing collusion. The effective power of these "arts of the underground" practiced by slaves and their descendants is not simply a point of scholarly fascination. It is revelatory of the character of modernity itself as a unique form of rationality that is "*actively associated with* the forms of terror legitimated by reference to the idea of 'race' " (Gilroy, 1993, 57; emphasis added). "The [modern] meanings of rationality, autonomy, reflection, subjectivity, and power" are not "merely compatible with" the practice of terror, but in some measure constituted by it (Gilroy, 1993, 56, 200). From a theoretical point of view, racial terrorism and modern rationalism must thus be thought together as profoundly *integral* to each other.

In his subsequent discussion of Richard Wright, Gilroy restates this perspective in a manner that clearly intersects with our own concerns for the connections between an incipient racism and a renascent soteriology. For Wright (as for Habermas and many other theorists) the fault line that defines modern Western consciousness was the breakdown of a cohesive religious worldview. But in *Outsider*, Wright positions this breakdown alongside a reading of blackness and the ideologies of racialization that posits the latter as part of the *metaphysical conditions of modern existence* that come into being precisely *in* this abandonment of religious morality. Wright's account underscores "the correspondences and connections which joined the everyday life-world of African-Americans *to the visceral anxieties* precipitated in modern European philosophy and letters by the collapse of religious sensibility in general and the experience of twentieth-century life in particular" (Gilroy, 1993, 160; emphasis added). Race and racism emerge here as integral to modernity as part of its peculiar metaphysical conditions of possibility. At the same time, that specific metaphysic is comprehended as dependent upon the very religiousness it repudiates, carrying it forward "viscerally" as a form of "philosophical anxiety." The religiousness is not entirely eclipsed by the new emphasis on philosophy; it rather goes underground, and continues to shape European subjectivity in the form of a repressed uncertainty.

On the other side of the divide, however—in the "unconscious" of modern rationality represented by modern slavery and its terrors—the denizens of darkness find the images of Jubilee and apocalypse specifically requisite for their survival. They deal with the quite physical anxieties of their situation in part by means of the very religious symbols their masters are in the process of discarding. While religion itself undergoes repression in European consciousness, it becomes the very language and force of explicit forms of subversion among those who are struggling against European oppression as its "others."

We could perhaps then summarize: the fears and figurations of a premodern soteriology show up with double effect on the *inside* of modernity. Against a double dismissal, they both irrupt and erupt. On the one hand, they recur symptomatically as a form of markedly European anxiety and terror. On the other, they find quite overt retrieval in the counter-forms of various black "transfigurations," directed precisely against the irrational projections and rationalized practices of that (European) terror. In the process, the various elements of that premodern concern-for-the-ultimate are reanimated with imprecise powers. In the radical self-splitting of a "rationalizing" Europe, those elements are simultaneously denied and displaced. They return as the histories of repression and oppression.

The modern connection between salvation and racialization, I would argue, must be grasped analytically in point and counterpoint. The complex duplicity of European racism confronts its other in the duplex complexity of racialized identity. It will find historical absolution only by way of a courageous encounter. Salvation may indeed be offered to all, but its way lies in paradox. In a broken-in-two modernity, the "logic" of salvation, as we shall see, must necessarily run through the historical nexus of that break. Modern soteriology speaks its word in the place of slavery and its overcoming. White anxiety will find its release only in facing the terror from which "whiteness" itself has arisen in the first place.

The Crisis of Race in the Theory of Black Resistance

As already indicated, the specific resistive powers of slaves and their descendants to the practices of modern racism will be covered in more detail in chapter 5. Here, it only remains to provide a theoretical framework for the analysis of one of their historical "children" in America. Obviously, the distance between nineteenth-century slave practices and black politics of the 1960s admits of much change and commands ever-renewed analysis. For our purposes, however, black intellectual appropriation of Hegel's master/slave dialectic can helpfully specify the practical soteriological theme encoded in both slave resistance and Black Power rebellion. This inchoate theme is part of the deep-structure of the split modern subject that has been continually repressed in white social supremacy and continually reasserted in black cultural urgency across generations and geographies for centuries now. Frantz Fanon underscores the stakes, in this modern struggle for identity, of a no-saying that risks death; Gilroy, from a different angle, highlights the soteriological ground and "utopian beyond" of that negation.

The Crisis in Liberty: Salvation and Recognition

Fanon appeals to Hegel primarily "diagnostically" to delineate a profound problem in the racialized world he inhabits. As a native-born Martiniquean psychiatrist, working in both the Caribbean and Francophone Algeria, he is increasingly exercised by the colonial pathology he keeps encountering in both his practice and his everyday existence. *Black Skin, White Masks*, written in 1952, is his *exposé* of the complex. While Fanon's overall strategy in that work is to project the possibility of future harmony for whites and blacks, he is adamant that the way forward lies not through mere analysis alone, but through battle and world-change. Hegel delineates a major difficulty in getting there.

The particular problem, for Fanon, is indeed psychological. The history of colonial slavery has perpetuated its structures in the post-emancipation psyches of both whites and blacks. Analysis *is* needed.

But the contours of those psychic structures—and the correctives they point toward—exhibit a genesis and a resolution that is finally *social* in origin (Fanon, 100, 104).

Appositely for Fanon, Hegel's master/slave allegory stages the requisite therapy in its very apology for the necessity of the problem. That allegory maps a particular "recognition-economy" that not only defines the modern predicament of race, but foreshadows what is implied in its overcoming. For Hegel, becoming human can only be achieved by way of forcing absolute recognition of oneself by another. Such recognition is the very meaning of human freedom and identity. In its achievement, the self attains certainty of itself as "a primal value transcending life" (Fanon, 218). But the drive for recognition is also the source of human conflict. Short of achieving recognition, the other remains a continuing theme of one's desire and action. Struggle continues, but without the possibility of a mutual resolution. In such an eventuality, consciousness takes shape as a form of unhappiness, unable to constitute itself in the truth of self-certainty.

For Fanon, much of this dialectic rings true. But what then of a form of recognition achieved *without a conflict*? What happens if recognition is won, but without passing through the absoluteness of the struggle in which life itself is risked? Here Hegel must be read against the grain.

The history of recognition with which Fanon is concerned—that of a unilateral white conferral of freedom unmaking slavery *from the white side only*—has arguably only continued enslavement in a confused form. In Fanon's opinion, Hegel was indeed right to insist upon the quest for absolute recognition as also a quest for absolute reciprocity of recognition. To achieve a genuinely human dignity of spirit, each consciousness must risk itself to the point of becoming a threat to the other. It must face into the other's fierce gaze, give birth to desire and plunge into what lies beyond life, risk death and meet the other's (reciprocal) resistance with utmost struggle. Without such a mutual face-off, Hegelian subjective certainty cannot transform itself into universally valid objective truth. To become a reality in-itself-for-itself, human "being" must pass through the phase of serving negative notice to its other: "Here I stake life itself; I refuse to remain immediately, 'naturally,' within myself."

But the actual history of emancipation Fanon has witnessed (at least in the Francophone world, up to the time he is writing) has not grown out of such reciprocity. It occurred rather as an upheaval within whiteness alone that changed neither black nor white, obliterated neither slaves nor masters. Coming from without, it effected no "difference within the Negro" (Fanon, 220). The latter created no value, constellated no true self-consciousness, memorialized no trace of struggle, in coming out of chains. Nor—in the new form of enslavement quietly inaugurated in the white pronouncement, "You are free"—did the "freed Negro" achieve independence by losing and finding self-consciousness in the object of work (the dialectical possibility inherent in the slave's situation made famous by Hegel) (Fanon, 221). Rather, Fanon recognizes only a "relapse." The liberative possibilities of objectifying consciousness in work have been foregone in a (re)turn to the master as a substitute "object" of emulation and desire.

In consequence, in its French colonial legacy, the black gaze (when Fanon is writing) cannot find its necessary opposition, its difference-that-constitutes, but only *in*difference, paternalism. It thus remains in search of the death-challenge. It is absorbed by the quest to uncover a difference it can fight for. It itches for "conflict, for a riot." Only when he looks to the United States, with its "twelve million black voices howling against the curtain of sky" and tearing it, does Fanon see hope—a hope born also through torn black bodies (Fanon, 222). Only there can he glimpse a "monument being wrought." To be sure, it is a monument of battle, but it is one topped by "white and black *hand in hand.*" Whereas the French black remains inured in an unpronounced rupture. . . .

The diagnosis is acute. Fanon sums up human being as a simultaneous "yes" and "no": a self, taking its place by wrestling with itself. Essentially, Fanon pronounces yes to life, no to every butchery of human freedom.

In subsequent colonial struggle, Fanon becomes *de rigour* reading. His Hegelian exposition of the impasse of post-emancipation racialization plots a course that Atlantic black nationalist and African independence movements alike will tread with unmixed fervor and very mixed results. The Black Power culmination of the Civil Rights movement in the United States will not prove an exception. Taking up the unfinished business of emancipation and pronouncing an emphatic no! with one's body (at whichever end of the gun it finds itself), marks the heart of the moment (as we shall see below). But at the same time (back in the text), Fanon's humanism exhibits the continuing problematic of the very course his Hegelianism has made irresistible. How to speak a "no" without simultaneously ramifying and thus perpetuating what is being negated? How to speak a "yes" without being shortchanged and seduced in what one thereby "receives"? How to avoid becoming—as he says in refusing the Sartrean vision of *negritude*—merely a "minor term in a dialectic"? (Fanon, 138).

Fanon's account is obviously richly evocative and is so, in part, because of its cadence, its almost guerrilla-like quality of having its say on the run, aphoristically. It provokes, rather than prescribes, and eludes precise solutions, pat equations. In its very manner of appropriating Hegel, reiterating his dialectic with a difference (in the post-emancipation slave who substitutes the master-object for the work-object, and in the master who seeks not recognition from that "freed" slave, but only more work)—Hegel ends up re-inscribed inside of Hegel in a sort of spiral. The very incompleteness of the analysis pointed out by Sartre—the fact that Hegel does not deal with lateral relations between masters, or those between slaves, or the impact of a free, non-slaving population on the slave institution itself—is only redoubled here by Fanon (Gilroy, 1993, 54). But within Fanon's multiplication and nuancing of the forms of slavery, there also emerges the demand for, and possibility of, different forms of no-saying.

The Crisis in Dignity: Salvation and Contention

Thus Fanon's text illuminates a set of questions *apropos* of our white approach to Black Power and Black Theology. How, within the terms set up here, could we assess, or more pertinently, even "recognize," where and when, within subaltern[31] histories of struggle, more hidden forms of "no-saying" had taken place? At what depth, in what overtness of public display, must a "no" participate to be risking life and thus raising self-certainty to the status of truth? Is there a form of minority or black-on-black recognition that takes place under the veil of a dominating whiteness, gaining its objective value precisely in the bet that it will (and thus also, the risk that it might not) *supersede* white comprehension and recognition? And what of whiteness itself in this equation, a whiteness that on Hegel's own terms cannot constitute itself except in the same risk, cannot unilaterally come to be in-itself-for-itself except through a struggle that it must *not* win?

Gilroy's similar redeployment of Hegel to illuminate Frederick Douglas's nineteenth-century account of his death-struggle with his master leads into further nuances. Within the structures of plantation slavery, the only forms of reciprocity available to the slave were "rebellion and suicide, flight and silent mourning" (Gilroy, 1993, 57). Only when slave Douglas resolves to stand up to "slave breaker" Covey (who was threatening to whip him), does their relationship suddenly exhibit "mutual respect." In a partially tongue-in-cheek reference to Habermas, Gilroy says, "Douglas discovered an ideal speech situation at the very moment in which he held his tormentor by the throat" (Fanon, 1993, 62). A two-hour standoff concludes with Covey giving up the contest and Douglas giving up his "spirit of cowering." In Douglas's own words about the incident, written years later, long after he had literally stolen himself away from slavery, he affirms, "I was nothing before; I was a man now . . . I had reached a point at which I was not afraid to die" (Douglas, 190).

But the point here is not merely that of a countersign to Hegel's claim that intersubjective dependency (i.e., "slavery") is a necessary precondition of modernity (Gilroy, 1993, 68). Rather, Douglas's words point to a broad range of instances of slave preferences for "death over bondage" that defy merely rational calculus. In Gilroy's analysis, this preference issues in full-blown acts of subjective agency—including acts of slave suicide and infanticide—that are often not evaluated as such by those who have not faced similar depths of desperation. In particular, Gilroy has in mind the example of Margaret Garner, whose foiled attempt to escape slavery in 1856 resulted in her opting to kill her three-year-old daughter with a knife (and trying to kill her other three children as well) rather than let her be taken back to slavery by her master.[32] Garner had reportedly told Lucy Stone, the white abolitionist who visited her in prison (and later described Margaret's motivation as "no wild desperation, but a calm determination"), that "she had made sure . . . her daughter would never suffer as she had."[33] In referring to this incident alongside of the one involving Frederick Douglas, Gilroy clearly wants to acknowledge gender differences in racial politics. He asks, without answering, "What are we to make of these contrasting forms of violence, one coded as male and outward, directed towards the oppressors, and the other, coded as female, somehow internal, channeled towards a parent's most precious and intimate objects of love, pride, and desire?" (Gilroy, 1993, 68, 66).

But Gilroy's point is not precisely that particular difference. He is more concerned for a wider set of meanings that were brought famously to the fore in an encounter between Frederick Douglas and Sojourner Truth that Gilroy introduces into his text. As recounted to W. E. B. Du Bois by Wendell Phillips, in a speaking engagement once in Fancuil Hall, Douglas had thunderously pronounced upon the lessons he had learned from his slave experience. "The Negro race," he had said, had "no hope of justice from whites, no possible hope except in their own right arms. It must come to blood! They must fight for themselves" (Du Bois, 1921, 176). In the "hush of feeling" after Douglas sat down, in the "deep, peculiar voice" that was her trademark, Sojourner Truth had then spoken out in a manner all could hear and asked: "Frederick, is God dead?" (Du Bois, 1921, 176). While Truth's trope here gives obvious evidence of a possible gender difference within slave politics, Gilroy does not use the incident to try to specify a connection between gendered practice and religious perception. He only notes that it is extant in the archives and bears investigation. He rather emphasizes that slave preferences for death over bondage—in *all* of their different forms—represent a distinctly *eschatological* "principle of negativity" (Gilroy, 1993, 68).

Whether in Douglas's overt resistance or in Garner's covert infanticide, in Douglas's war cry or in Sojourner Truth's outcry of faith—what is at work in slave counterviolence is a form of black spirituality focused on the moment of jubilee. Its discourse "possesses a utopian truth content that projects beyond the limits of the present" (Gilroy, 1993, 68). This is more than mere dialectic, more than the rational calculation typical of Western modalities of logic. Even Douglas's "distinctly masculinist resolution of slavery's inner oppositions" is animated by his "skeptically Christian" religious convictions (1993, 59, 64). Interestingly, Gilroy fails to mention that Sojourner Truth's moment of "signifying on" Douglas met with a less well-known rejoinder from him: after a pause, he said, "No, dear sister, God is not dead, and because God is not dead slavery can only end in blood."[34] But he does note that Douglas himself had been given a " root amulet" by a conjurer the night before his standoff with Covey—an African traditional power whose efficacy he could neither entirely believe nor entirely discard. Garner, on the other hand, had less room to maneuver, and could only offer, in support of her actions, a grim determination. As Stone recounts her saying: "if she could not find freedom here, she would get it with the angels . . . but she had made sure" that her daughter would find it sooner rather than later (Blackwell, 183–184). And Sojourner Truth's fury is perhaps best heard not so much as a comeback to Douglas as a comeuppance addressed to the culture at large: the cry of a black apocalypse, standing Nietzschean-like perceptions of "God's death" on their head and asking if indeed modernity has not rather killed its God. As we have said above, and will look at below, the point here is a facing of terror that effaces white pretension to playing god. The choice for death is an exorcism. For the community that passes through its waters, and remembers its passage, salvation has a face of dread hope[35] that rationality can never see.

Thus we can summarize. Fanon, by saying the unsaid in Hegel, focuses the task of a black post-emancipation politics on a no-saying that embraces and moves beyond its own death. Gilroy, in an inversion of the priority of philosophy, returns religious depth to the Hegelian absolute—or more precisely, returns the Hegelian dialectic to the terrors, and triumphs in that terror, of a "black" absolute it cannot simply synthesize. But it is Hegel himself, who remains the voice of questioning for a white theology watching all this unraveling and reknitting. What is such a theology able to recognize? What would it take for it to recognize *itself* in a living process—not merely a dead dialectic—in which black and white "recognized themselves as mutually recognizing each other"? (Hegel, 230). Here is the rub of the Black Power uprising that, for a brief historical moment, looked its other in the eye and met no light of recognition, but only a fear born of guilt. Black Power raised the still unrealized question of a necessary *white* no-saying, different in kind, but no less deep in death and terror. Strangely, it is this "depth of death" that is so easily missed, so readily dismissed, in the very theological-ness of a Black Theology thinking "Black Power." For I also want to argue, as others have before me, that "theological-ness" itself, at least in its academic forms, is already a matter of playing on the institutional terrain of a white drive for certainty and control. I would only add: it is a drive for certainty that has not yet passed through the baptismal depths of a racial death and risen to itself in truth. If so, Black Power bears scrutinizing for a moment absent its Black Theology "husk."

The Crisis in Society: Salvation as Black Power Exhibition

The 1960s emergence of the Black Power movement theologized by James Cone as a public "epiphany" of blackness marked a new moment of self-revelation for modern black cultures at large and a new boldness of counter-racist combat. In the global context, it partook of a broad wave of anticolonial nationalisms surfacing when the industrial phase of capitalism was at its zenith. Closer to home, it found some precedent in early twentieth-century Garveyism and the later rise of the Nation of Islam. It did not leap, full-grown, out of the head of the Student Nonviolent Coordinating Committee (SNCC). Black militancy had a history. But for the most part, until the mid-1960s, African Americans had been reticent about asserting their blackness on a public par with the whiteness that constituted the normative content of "American" identity. Most public gestures had appealed to an integrationist sameness, not an explicit and diffident difference. No-saying had largely been conducted in the forms of a rational political discourse defined by white law and culture. Never had its content been broadly and unabashedly "black."

The reluctance was, indeed, born of a long history of painful public experience. In the post–Civil War south, reconstructionist hopes had been crushed by the politics of disenfranchisement. From the late nineteenth century onward until the Civil Rights movement, legalized Jim Crowism and legally unrequited lynching monitored the public square for any shows of "upitty-ness." In the north, the less obvious, but no less onerous, forms of racist practice, such as social enghettoization and employment discrimination, similarly eroded the hopes and trimmed the hackles of those who immigrated from the south. As works like Richard Wright's *Black Boy* and *Native Son*, or Ralph Ellison's *Invisible Man*, made so graphically clear, black life, both south and north in the first half of the twentieth century, was continuously vulnerable to violation and demanded ever-adaptive skills of clairvoyant anticipation, canny self-invention, and careful negotiation. Blackness "militated" itself only covertly.

That covert-ness was uncovered without apology from 1966, on, however. Whether conjured in the clenched-fist-forcefulness of a Stokely Carmichael or called out in the razor-sharp-rhetoric of an H. Rap Brown, whether paraded in the black-beret-ed, black-jacketed, black-gun-toting of the Black Panthers or preached in the black-Jesus worshipping, black-Mary reverencing, black nation-building of Albert Cleage and Detroit's Shrine of the Black Madonna—blackness had a new presence in America. And the new heralds of that presence were only following the long suit of Malcolm X's fearlessness before them. Malcolm had made his mark—and been made to bear the price of it—not only in what he said, but in the way he said it (Cone, 1991, 304–305, 310–311). Not only acute analysis, but also acerbic eloquence, figured in his impact. In the steely-ness of an eye and the thrust of a chin, black consciousness had begun to measure out a new meaning. It filled up its given quota of public time and space—or, as in James Foreman's foray, created it for itself, where it was not given—with its own chosen forms. And its most often chosen form now was the style of *power*—even if its legacy was destined to come down to the present as largely the power of "style."

The moment was quintessential in the history of modern racialization. The grammar of control had been broken apart and repunctuated with different "breaks," different pauses and timing. Words themselves were drained of connotations and refilled with the denotations of a community coming out of its centuries-shut closet. "Blackness" remained a ubiquitously operative signifier in the discourse of race; now it mattered who was using it, where, and with what gestures on whose stage. The reformulations of public blackness on black tongues figured the conflict of the country at large; its various contents hosted the history. Black Power as a movement made frighteningly unavoidable for whites what had always been painfully unlivable for blacks: the ominous eventfulness, and simultaneous vacuousness, of the category of race. It postured blackness in a public meaning never before staged quite that way: defiant, impenetrable, death-defying when necessary, life-threatening if necessary. Recognition of this assertive challenge constitutes one of the poles of black significance for an interested, or "convicted," white theology.

But the force of Black Power was not merely negative. Its "visage of terror" for whites was not only that of an opposition, a reverse sameness. Rather, in itself, in its various displays and claims, Black Power could be said to have briefly publicized an alternative mode of inscrutable being. Among other things, it constituted a popular rejoinder to the autonomies and antinomies of modernity. The historic black agony that Civil Rights rhetoric and dramaturgy staged in the public square of America as a kind of "homeopathic remedy" for racism,[36] blew wide open in the decade following 1966. Black assertiveness broached a chasm in the consciousness of the country at large. What surfaced was the return of the repressed—precisely as the tactics of the oppressed. Militant blackness, understanding itself as a spiritual counterforce[37] and organized as a political practice, perhaps cohered most effectively as a cultural aesthetics in the Black Arts Movement of the period. As in earlier modalities of expression, this black cultural production refused, in practice, the split between art and life, ethics and aesthetics, poetics and politics, characteristic and constitutive of Western modernity *in toto*.

Black consciousness of the late 1960s emerged as a form of self-styled "power"—actually, a counterpower to the powers of racialization and racism it both reflected and transcended. As power, public blackness addressed itself simultaneously to the white structures it contended against and to the black communities it sought to speak for. In this latter capacity, the movement undertook and underwent various reconfigurations. Black Power's "I'm black, I'm proud" voice articulated a *counter-ethics* of assertion in countless public litanies and grassroots organizational pedagogies over the course of subsequent years. "Black is beautiful" incantations invoked a *counter-aesthetics*, exploring a new natural style that circulated widely in subsequent African American graphics and performance productions. The grunts and groans, slides and syncopations of a James Brown or an Aretha Franklin launched a broad *counter-spirituality*, retooling black gospel and spirituals as a new and secular form of "soul" music.

At the same time, militant blackness remained as ambivalent as any other claim to or expression of power. It could be mobilized ethically for ill as well as good—as any accounting of the movement from the point of view of gender relations and sexual orientation (or perhaps even class analysis) would show. It could show its face aesthetically as beautiful . . . or less than beautiful. It could further life, spiritually, or threaten death, play host to the divine or dance close to the demonic. Its aspect of alarm for whites lay not so much in the forceful-ness evident in any one of these arenas. It lay rather in the interlinkage of all of these with its political meaning as death-defying.

Blackness in the Black Power movement signified an absolute no-saying that equally embodied an absolute yes. The "yes" was spo-ken in various modes of communal practice that projected, even if they did not realize, an identifiable counter-coherence to the incoherence regularly experienced in wider society. At a deep psychic level, they tattooed public awareness with a recurrent image and partial practice of a utopic aspiration. The black way of being-in-the-world thus exhibited did not only terrify, but finally, also, mysteriously attracted. For the first time ever in American history—in the full glare of the media gaze—black identity was displayed as the proud sign of a mani-fold ability. Before the uncomprehending gaze, but not unfeeling gut, of a cowed and angry white public, black style asserted a double dare, a twofold trope.

In the presence of this newly assertive public blackness, whiteness was "outed" and light-skinned people were left fumbling with both fear and fantasy, guilt and grace, damnation and salvation. In the very excesses of its repressive responses—the more than 80 bullets fired, for instance, at the sleeping body of Black Panther Fred Hampton and his housemates—the dominant white order belied its force as rational, as a merely instrumental "execution" of the law. The responses were rather ritual—cultic actions of mythic proportion, overdetermined intentions of alarmed awfulness. At some profound level of inchoate desire, perhaps envy even, white people "knew" themselves to be doubly displaced. Even uncomprehending, they registered an image and felt a practice that had not capitulated—to the same degree most of them had—to modernity's disciplines of the body and its strictures of soul. In a vague instinct, "whiteness" both recoiled from and dissembled toward a position not merely opposite, but tangent, to its own fragmentary-ness. White people were finally, briefly, made to look in the mirror of race and confront a gaze that looked back, but did not look alike.

It is not my intention here to construct a panegyric to the Black Power movement of the 1960s, or to its Black Arts offspring of the 1970s. (In part, the Black Power phase of the freedom movement only served to further divide the urban black community along class lines. Initially reflecting, in the passion of its style, the ferocity of the bottom levels of the social structure, Black Power was effectively "de-truncated" as a movement in the "cooptation" of some of its top leaders into the few white institutional openings won in various forms of negotiated settlement.) My purpose here is rather a matter of securing the (late) recognition of a moment of epiphany, of the appearance of something that had been practically perduring since modernity's inception, but had enjoyed scant moments of public expression and even scanter moments of toleration or comprehension in the white public sphere. It was a significant moment in the racialized complexity of American modernity that demanded and continues to demand adequate signification. In the realm of theology, it found the beginning of such in the work of James Cone.

4

Black Double-Consciousness and White Double Takes

Ever have men striven to conceive of their victims as different from the victors, endlessly different, in soul and blood, strength and cunning, race and lineage. It has been left, however, to Europe and modern days to discover the eternal worldwide mark of meanness—color!

—W. E. B. DuBois (The Seventh Son, 495)

When the video-taped beating of Rodney King by 21 officers of the law in California was shown on newscasts around the country in the days following that March 1, 1991 debacle, a private transcript of white America was suddenly rendered public. Racialization was graphically displayed in one of its most virulent forms. An African American male body was "made black" in the action of 56 baton blows that spoke volumes in a spectacle almost devoid of speech. In many respects, the act was archetypal. White paranoia could be observed, freeze-frame by freeze-frame, creating its fantastic other. In an astonishing displacement of aggression, armed white authority recast a supine, elbow-in-the-air-palm-turned-outwards-to-deflect-the-blows, dark body as "bestial," a "gorilla in the mist." In the same instant and action, it postured itself as "imperiled," supposedly under threat of the terror that, in fact, it had itself authored. Here for all to see was American order producing its necessary criminal in the very act of policing. Arguably, the excess of that moment remains unassimilated in the country at large. Its overdetermination as a moment when many discourses, many histories, many institutional powers, and many cultural processes fused violently together became clear only in its explosive aftermath: the social "fission" of South Central, Los Angeles from April 30 through May 3, 1992. The displacement of white violence condensed into a "nuclear" black and brown and yellow and tan and white reaction. But it was a chain reaction that found its genesis in a split-second perception. The moment of being incarcerated in blackness by a white eyeball—with all the violence that attends that momentary vision—has perhaps found no more compelling exposition than in W. E. B. Du Bois's description of the experience in *The Souls of Black Folks* or more damming elaboration than in his *The Souls of White Folks*.

Why Du Bois?

Chapter 3 traced the historical rise of racial supremacy as the modern offspring of a form of "theological" supremacy that first emerged from Christian–Muslim encounter in medieval Spain, was ramified by colonial encounters with other cultures and peoples in the Americas, Africa, and Asia, was given philosophical and "scientific" underpinning during the Enlightenment, and now is globalized as the "hidden transcript"[1] of a world system brokering resources and power across the planet. The race discourses projecting such a supremacy were exposed as a *de facto* form of soteriology, delineating the social map of entitlement and access, both materially and spiritually, inside the entire project of modern development and operating as a kind of perceptual shorthand for a depth-structure or subtext of the modern Western subject, marking the boundaries and character of taken-for-granted notions of wholeness. Black Power was read as embodying a counter-soteriology, functioning like a return of the repressed from the days of slavery, outing white supremacy and confronting America with an alternative orientation toward wholeness. The ensuing battle of mythologies—of white supremacy and black soulfulness, of ownership claimed and denigration reframed, of heaven assumed and hell exhumed—points toward a common structure of identity in America even as it marks out the terrain of an indelible difference. This chapter initiates an exploration of that double bind by way of W. E. B. Du Bois's divination of the souls of black and white America from inside the body that has borne the pain of a split demand. Black skin, in Du Bois's light, is a palimpsest of the writing of race, even as black consciousness is the penetration of color's blindness. What blackness is aware of, in the history of America, is the subject of chapters 5 and 6.

Postscript on a Struggle: Du Bois and the Production of Blackness

This chapter turns away from the broad sweep of history to the close inspection of phenomenology to track the way myth marks its contradictions in the flesh by means of gestures minute and metabolic. As we shall see, in matters of race, an eyeball can eat (substance) even as it incarcerates (soul). Du Bois's writing will provide the paradigm that reveals the phenomenon of supremacy at its inception and of racialization in operation. A close reading of his description of the experience of being racialized and rejected will exhibit the subtleties of the violence involved while simultaneously demonstrating a way of talking about it that resists repeating the reification "in kind." The double-consciousness formula that Du Bois crafts as his cipher for such an experience offers a very productive metaphor for theorizing the dilemma. But it also points to the difficulty. It forces recognition that racial struggle takes place not only between, *but also within*, individuals and communities. The fact that Du Bois's delineation has become so widely used only underscores the breadth and complexity of the problematic it glosses: America at large, indeed, modernity itself, are racialized constructs that are not simply exorcized in being so named.

White experience as well, however, is signified (and signified upon) in Du Bois's formula and later elaborated in naked display in his rampage on the white soul. If "being made black" is an experience of a kind of "hyper-consciousness"—a doubling of awareness that distends and dilates—we can also burlesque Du Bois's formula to caricature white identity as a condition of "double-*un*consciousness." At one level, whiteness is simply the unself-conscious reflex of a perception of difference that racial discourse and social structure stigmatize as "nonwhite" and "inferior." At another level, it is the unexamined presumption of entitlement to material wherewithal gathered historically and continually from the populations of color that have been marked and constrained by those discourses and structures. White identity is generally lived, that is to say, as a kind of "artless ignorance," an almost incorrigible lack of awareness of either one's racial position or of the actual cost to others of one's prosperity. Du Bois's way of talking begins to lift the veil of white ignorance precisely in describing the way the veil of black difference descends upon his own consciousness. "Black" and "white" in his discussion are anything but black or white.

But there is also a polymorphism of the pun at work in Du Bois's language here. Being "born with a veil," as Du Bois will describe it, is actually a "veiled" reference to a revelatory birth experience in black folk tradition—the occasional appearance at parturition of a kind of filmy gauze on the infant eye, signifying a gifting with "second sight." The premonition encoded in the appearance of such a "caul" (of filmy mucus) is a propensity for shamanic potency in perceiving the violent undercurrents of everyday life—the writhing exchanges of desire and denigration locked in titanic struggle like ancient gods under the surface of "polite conversation." The veiled eye is a voodoo iris, looking straight through the lie. And there is also arguably an apocalyptic play in the word choice, a subtle solicitation of the Book of Revelation (literally, The "Apocalypse" or "Unveiling" to the Apostle John), in the sudden appearance, to the eye of blackness, of this veil that masks. What is masked, Du Bois will reveal in *Souls of White Folk,* is a beast of prey of apocalyptic proportions. However subtle its everyday operations, white supremacy is ultimately a force of imperial ferocity—Gorgon-like in its capacity to mesmerize while consuming a world. Du Bois "unveils" the veil of race as cold ideology.

Thus in Du Bois we get a black "sounding" of the meaning of white identity in both its everyday prosaic-ness and its ultimate consequence. If, as we discussed in chapter 3, soteriology is the Christian discourse that most clearly articulates the stakes at stake in racialization, it is apocalyptic imagery and mythic symbology that most adequately augur the depths of the dilemma that must be faced. Du Bois opens our exploration of the black depths of white supremacy in the perspective of phenomenology. Chapter 5 extends that exploration in the key of cosmology and mythic struggle.

Phenomenology of the Experience of Racialization

Du Bois's description of racialization as the "dropping of a veil" in the first chapter of *Souls* augurs the event that sets his life course and serves as touchstone for reflection here.[2] Before articulating his (in) famous formula, he outlines a formative *experience.* The animation and forcefulness of the long career of social combat that his life came to embody actually grew out of an anguish that began much earlier and germinated much deeper than any of his written words would ever succeed in fully conveying. (Even in that first writing, there are already

hints of the incapacity of the tongue to mollify the *pathos* of the body.)
"It is in the early days of rollicking boyhood," writes Du Bois, "that
the revelation first bursts upon one, all in a day, as it were" (Du Bois,
1961, 16). And immediately, he shifts voice, "I remember well when
the shadow first swept across me." Only a few sentences later, he has
migrated again: ". . . The shades of the prison-house closed round
about us all . . . walls straight and stubborn . . . relentlessly narrow,
tall, and un-scalable to sons of night who must plod darkly on in resignation, or beat unavailing palms against the stone, or steadily, half-
hopelessly, watch the streak of blue above" (Du Bois, 1961, 16).

Skin as Script: The Genesis of Struggle

In a single long paragraph, whose blanks of early experience we shall
fill in below, there is a rapid shifting of pronouns from the third person
singular ("one") to the first person singular ("me") to the first person
plural ("us"). Du Bois writes of his boyhood revelation in the forms of
a subjectivity that shuttles constantly between the individual and the
social, between the past and the present, indeed, between himself as
subject and himself as object. His anguish is rooted in a collective
experience and a historic memory that remains plastic and prolix and
plural. In his community, there was more querying of God, more plod-
ding of the feet, more beating of the palms, more watching in despair
than the words of any unitary self could ever account for.

The allegorical quality with which Du Bois invests this childhood
reminiscence make its words worth quoting and pondering at some
length at the outset of this chapter.

Prescript of the Struggle: The Priority of a Glance

The first two pages of *The Souls of Black Folk* probe Du Bois's
childhood memory poetically:

> I remember well when the shadow first swept across me. I was a little
> thing, away up in the hills of New England, where the dark Housatonic
> winds between Hoosac and Taghkanic to the sea. In a wee wooden
> schoolhouse, something put it into the boy's and girls' heads to buy
> gorgeous visiting-cards—ten cents a package—and exchange. The
> exchange was merry, till one girl a tall newcomer, refused my card,—
> refused it peremptorily, with a glance. Then it dawned upon me with
> a certain suddenness that I was different from the others; or like, may-
> hap, in heart and life and longing, but shut out from their world by a
> vast veil. I had thereafter no desire to tear down the veil, to creep

through; I held all beyond it in common contempt, and lived above it in a region of blue sky and great wandering shadows. That sky was bluest when I could beat my mates at examination-time, or beat them at a foot-race, or even beat their stringy heads. Alas, with the years all this fine contempt began to fade; for the worlds I longed for, and all their dazzling opportunities, were theirs, not mine. (Du Bois, 1961, 16)

It is not hard to read paradise into this picture, idyllic and soft—until the peremptory-ness intervenes as small as an eye, as sharp as light. Nor, it is worth noting, is any color mentioned—or rather, it is a world full of wonderful color, "innocent" darkness . . . until it coagulates, and is made to fall, like a veil. But DuBois's characterization of this sudden advent is strange, paradoxical. His first description of it is as a "shadow" that "swept." Then, a scant four sentences later, it is a "dawn." It "dawns" *up*on him . . . (from on high, gift of a "tall new-comer"). Indeed, much later in life, he will still be waiting for the light—writing of that moment, of the genesis of that day, and the days before and since, as the "dusk of dawn" (the title of his 1940 book).

And perhaps it is no mere fortuity, no accident of incident, that the writing frames the veil-falling in the language of "exchange"—a dime for a dime, seemingly. There is a refusal (of Du Bois's gift), but with a substitute in place, indeed, two substitutes. Rather than a card, Du Bois is offered peremptoriness, in a glance. The beginning of the rupture that carves Du Bois's world into two discrete ones—two life-worlds, lifelong in their juxtaposition—is a single look: the power of a look. Du Bois's choice of words repeats the economy of the moment. It was, Du Bois is explicit, a peremptory look, literally, a look that "closes out" (from the Latin, *emptum*, as also in "preempt," to "pur-chase before others"). The word is noted by Webster as a "conclusive, an 'absolute,' taking away of a right of action or debate"; it is noted also, with its qualification *per* in the Latin, as "destructive." It is a look that buys—we are not preempted here from reading Du Bois closely, historically, subtly—a look that has so often bought, or at least thought to, in this history not yet concluded (in spite of all its "eman-cipations," its moments of "rights" and "affirmative actions," since 1863). A look that buys and takes away and destroys all in the open-ing of a lid, perhaps the cock of a head, a certain gesture of the face.

Du Bois is canny in his description of this uncannyness, this sudden-ness of a difference that sticks to him, even as he still thinks in terms of likeness, stuttering "or like, mayhap" (". . . perhaps . . . possibly . . . maybe . . ."). But from now on there is a "but" slicing the likeness: "I was . . . like, mayhap, . . . *but* shut out." The likeness was cut open, not with a word, but with that something that appears in the eye.[3] The genesis of difference that we as readers can observe in Du Bois's words—the flow of images that come one after another as a slow dawning, letting us also play the awareness, the insight, in reverse, go back and reread and fathom its emergence—is not such for the eager child. For little Du Bois, in the moment at hand, the experience is rather an irreversible "worlding event," the birthing of a second world from the ripped body of the first one, primordial, like so many myths have it (in speaking of the world as so many body parts of a defeated god or goddess). It is the sudden, momentous intimation that the body of one's "world"—of one's cosmos and context and consciousness—has been ripped, ruptured in a misty beginning (for some infraction, some "something" gone wrong), unaccountably forced to give life to an other world by dismemberment.

Do I overweigh the passage? Perhaps. Yet it is arguably Du Bois's point, as we shall see as we go on, and see with much more scholarly apparatus, more justificatory logic. The "color line," the "great problem of this century," the immediate excuse for all the wars of the "comfortable civilized" against the "impoverished majority" that so exercises Du Bois's pen and animates his life, until it is world-weary and resigned, is born in a glance (Du Bois, 1961, xiv, 16). Du Bois does not even assign the experience a "color" in the moment of its epiphany. It is simply, irrefragably, irremovably the experience of the eruption of difference: "Then it dawned upon me with a certain suddenness that I was different . . . ." Or perhaps, we do better to locate the moment in the word "from"—"I was different *from* the others." The sublime, the terrifying advent of a new god, a "between" god, a force unannounced, unheard, even invisible—although the coming happens in the eye, through the eyeball and is received by such. The great god *From*, rebirthing Du Bois himself, giving him a new point of origin, a new parentage, a new womb. From now on, he is "from" the others. Or at least such is the uncannyness of the moment of racialization, experienced on the other side of the one who racializes.

But another question also wants asking here, before we get on with our logic, our setting up of the text for analysis. How do we know it is race that is in question? In the one scant paragraph that precedes this one in his text, Du Bois lets us know that he is speaking about skin color only in the barest of hints. By and large, he has evoked that interpretation—irrefutably the *right* one—from our own eyes as readers. He has "made" us read color into the text from our own history, from its aliveness under our own feet, in our own "text." In understanding

the very meaning of what Du Bois is writing—so prosaic, so clear at
face value—we have to repeat the very thing he is describing. We *see*
race "into" the text; we produce it, full-blown from our own eyes.
And in so doing, *if* we choose to see, we can see that our eyes are not
innocent. It is this moment of genesis and this production that I am
interested in attending to, in what follows, as *the* question that must
be engaged by white conscience. It is at one and the same time, the
genesis of blackness *and of whiteness*. But the latter happens only by
way of surreption.

Scripting the Glance: Double-Consciousness as the Failure of Recognition

But we are not ready to engage such yet. The account Du Bois offers is
paradigmatic. It represents a kind of experience that shows up in
numerous other African American writings (e.g., in Fanon's *Black
Skins, White Masks;* in Patricia Williams's *The Alchemy of Race and
Rights*). It roots racialization in a split-second occurrence, freighted
with history, with memory, with meaning. But for all that, it is not
necessarily a moment born of intention; it is not necessarily "adult," it
is not even, we could perhaps argue, "conscious." But it does produce
consciousness. In Du Bois's text, only three more sentences—
describing the struggles of blacks to survive in the "prison-house [that
has] closed round about us all"—separate this experience from his
most famous passage (Du Bois, 1961, 16).

> After the Egyptian and Indian, the Greek and Roman, the Teuton and
> Mongolian, the Negro is a sort of seventh-son, born with a veil, and
> gifted with second-sight in this American world,—a world which yields
> him no true self-consciousness, but only lets him see himself through the
> revelation of the other world. It is a peculiar sensation, this double-
> consciousness, this sense of always looking at one's self through the eyes
> of others, of measuring one's soul by the tape of a world that looks on in
> amused contempt and pity. One ever feels his twoness,—an American, a
> Negro; two souls, two thoughts, two unrecognized strivings; two war-
> ring ideals in one dark body, whose dogged strength alone keeps it from
> being torn asunder. (1961, 16–17)

The product of such a gaze of race as Du Bois has traced to and
tracked from his childhood is a consciousness that is double. Its characterization in this passage is quite dense and overlaid with intertex-
tual reference. It warrants close reading.

In Du Bois's language of the "seventh-born" we are immediately transported into Hegelian discourse, witnessing a typically Du Boisian tactic in struggle. Du Bois parades—with intimate familiarity, without announcement or fanfare—his "American" erudition, his consorting with European thinkers. He is no captive to the gaze that has captured him, staying docilely in place, at a distance. He does not disdain "white" theory; he has instead wrested its prizes to himself. Here, he speaks the language of philosophy of spirit, uses Hegel, in order to speak within, but beyond, Hegel himself. He opens Hegel's discourse to a difference it did not anticipate, as Fanon and Gilroy were to do after him.

Du Bois actually begins the double-consciousness passage with Hegel's philosophy of history, inserting, into the Hegelian list of six world-historical peoples, "the Negro" as "a sort of seventh" type, but "veiled," and in the American world, "gifted." But quickly, then, with not even a period intervening, Du Bois shifts over into the phenomenological questions of the Hegelian master/slave dialectic. It is the recognition-economy we find outlined in that interminable struggle that provides the implicit background for the focus on self-consciousness here. The second-sight with which Du Bois "gifts" the Negro in the Hegelian list, is immediately overturned in its American context. For this latter world yields the Negro "no true self-consciousness," but only a seeing *through* the revelation of another. Double-consciousness, in its American provenance, is a "sense of always looking" to see how one is being seen, and then measuring oneself in the mirror of that seeing. And what looks on, does so in amusement, with either contempt or pity, as a looking that looks down. It is a looking that yields no Hegelian truth, but rather a particular American Negro feeling.

Double-*consciousness*, for Du Bois, is primarily a *feeling* of doubleness. It does not stay merely at the level of recognition, of gaze, but rather issues in a "peculiar sensation." The economy of DuBois's language here should not be allowed to eclipse the density of his figuration. It is a feeling that he describes in three ways. He offers first the juxtaposition of a national characterization and a racial one, with no qualification, but only a comma intervening: "one ever feels [one's] twoness,—an American, a Negro;" with the semi-colon providing a halt before the next characterization. This first articulation is then followed immediately by a shift to the language of interiority: one feels (more simply, more symmetrically) "two souls, two thoughts, two unreconciled strivings;" again with a semi-colon, but with the latter pairing beginning to require qualification as, somehow, "unreconciled," "at odds," "un-harmonized." Across another scant semi-colon, Du Bois's thought breaks into a still different characterization, this third time in terms of out-and-out combat, "philosophical" in nature, between "two warring ideals." And most significantly, this last metaphor is rooted finally in the *body*—for Du Bois, a "dark" corporeality whose strength of persistence, of "doggedness," alone prevents the doubled metaphysics from rending the physical reality itself. In a quick, poetic progression, we have thus shifted from a Hegelian economy of consciousnesses through a feeling-structure of nation, race, and psychology into a corporealization of idealisms. Or said more simply, we track the power of a seeing that affects a moving. Du Boisian double-consciousness is finally a body-structure of agonizing—and competing—intensities.

For Du Bois, double-consciousness is both affliction and resource.[4] He goes on to give a history of this strife as a history of "powers wasted," of "genius dispersed" (1961, 17). In its light he can explain much—the struggle of the black artisan, the temptation of the black doctor or minister, the conundrum of the black intellectual, the confusion of the black artist, even the clamorous celebration of "the ruder souls of his people" (1961, 17–18). Always there is this contradiction of "double aims," a continuing curse of being "half-named" but doubly claimed in a stream of prejudice that will not cease (Du Bois, 1961, 20). But at heart, Du Bois says, "the American Negro" seeks neither to "Africanize America" nor "bleach [the] Negro soul in a flood of white Americanism," but rather to "merge [the] double self into a better and truer self" without losing either of the older ones (Du Bois, 1961, 17). The aspiration is not a matter of eclipse, but integration.

But however the history is understood, existentially there is no let-up. Du Bois's very first paragraph, opening not only his writing on childhood and his double-consciousness discourse, but also the entire work of *Souls*, has begun the book in the present tense.

> Between me and the other world there is ever an unasked question: unasked by some through feelings of delicacy; by others through the difficulty of rightly framing it. All, nevertheless, flutter round it. They approach me in a half-hesitant sort of way, eye me curiously or compassionately, and then, instead of saying directly, How does it feel to be a problem? they say, I know an excellent colored man in my town; or, I fought at Mechanicsville; or, Do not these Southern outrages make your blood boil? At these I smile, or am interested, or reduce the boiling to a simmer, as the occasion may require. To the real question, How does it feel to be a problem? I answer seldom a word. (Du Bois, 1961, 15)

Between himself and the other world, Du Bois says, "there is ever an unasked question" that is "fluttered around" in indirect comments, "half-hesitantly" broached by an eye, curious or compassionate even, but never directly said. Even in adulthood, the eye remains the key for him. To that unasked question: "How does it feel to be a problem?" he answers "seldom a word." A whole history continues to be enacted—and repeated—below the surface of what is not said. That unasked question is the womb of whiteness.

Glancing Back Through the Script: What Double-Consciousness is Conscious Of

Writing 20 years later in *Darkwater*, Du Bois has only sharpened his insights into the unspoken depths. From a refuge high in the tower of academe, he looks across the wild waters of the world, fascinated above all by one wanton appearance: Whiteness. The souls of white folks ever vex and captivate his eye. Of them, he says, he is "singularly clairvoyant" (Du Bois, 1971, 486). He sees "in and through them," to the "entrails." It is a shaman-sight—the inheritance of the veil realized, the peculiar sensation, as we have already noted of double-consciousness, which augurs vocation and vision, possession and pain, under the regime of spirit. Du Bois's identification of second-sight as gift as well as affliction in his first book has borne through. "Bone of their thought" and "flesh of their language" as he remarks of himself, he does not come to his sight as a "foreigner": he sees white folks from within, "naked" (Du Bois, 1971, 486–487). And bears their wrath as result. The vision is "ugly, human," he says. But perhaps the only vein of self-realization in which white fury could ever find salvation.

Du Bois knows well the fury. It is for him, a modern discovery—this "wonderful" whiteness, ghosting the globe with its tirades of self-importance, its certainty of rightness. It sings to Du Bois, as indeed it had 20 years earlier, an obbligato tone above all its fine words, a sweet rag, full of pity and promise, handed down to the poor, un-white world, as hope of one day being born again on high—through lowly labor—as white! Du Bois does not laugh. He only asks, "But what on earth is whiteness that one should so desire it?" And he is then "always, somehow, some way, silently but clearly . . . given to understand that *whiteness is the ownership of the earth* forever and ever, Amen!" (Du Bois, 1971, 487; emphasis added). Du Bois elaborates an entire character from the tunes above, the tones below, the unspoken traces between the words. Title to the known world is the unwritten score.

But Du Bois is determined to write back. The pallor, he says, is wanton in comedy, strutting, running amok, whooping with English arrogance, southern mobbishness, hoodlum garishness, circling the wagons (Du Bois, 1971, 487–488). Whiteness waxes funny to the point of tragedy, claiming every attainment, every good deed or sober invention to itself, as if never a thing of value were created by a skin nonwhite. But when the hand-me-downs of self-invented *noblesse oblige* arouse humble wrath—when black pride disputes the white title, refuses the charity, claims a higher wage total, angles for position and authority—the unconscious spell of light is suddenly broken. Hell is not far behind. Du Bois names the discharge hatred—a bilious billow of barbarism and cruelty that targets poor black women, lost children, anonymous folk in the wrong Pullman car or waiting room or bus seat (Du Bois, 1971, 489). Hatred! . . . mixed with joy at the continuing misfortune below. Ironically, it is a thing toward which Du Bois himself—despite his own share of the suffering and "shackled anger" and blazing craziness of fantasized retribution—experiences pity. "Pity," he says, "for a people imprisoned and enthralled, hampered and made miserable for such a cause, for such a phantas[m]" (Du Bois, 1971, 489). White enmity provokes black pity.

Du Bois also names this white delirium "theory"—a tenet of morality, an iron rule of practicality, around which "defense of right" nationwide gathers and rages as soon as any violation whispers a mere drop of black blood (Du Bois, 1971, 490). Murder, theft, prostitution he names as all of "lukewarm" provocation (to the ordinary white person) until the perpetrator appears dark: in crime it is blackness rather than illicitness that is most condemned. And indeed, he will say, this rule is more than mere morality, but itself a religion, a wave, sweeping every shore now, teaching, as its new divinity, trade. It is a "theory" ranging far beyond national boundary, taking the world of color as its tenement, a house of gold to be divided up; its denizens worked, and warred, into European profit. America, on the scene, is a latecomer to the splendor, but now in the vanguard of raising up this peculiar "religion" as a "world war-cry: Up white, down black; to your tents, O white folk, and world war with black and parti-colored mongrel beasts!" (Du Bois, 1971, 500). Perhaps Du Bois's most prescient augury is the analysis of the first "War-to-End-all-Wars" as an intra-white dogfight over black spoils.

The 1915 turning of Europe from its self-imposed duty of terrorizing and pillaging the world of black, brown, and yellow (in the name of trade) to focus temporarily on killing each other, Du Bois and other

"Darker Peoples" watched with amazement. But not surprise. "Never," says Du Bois, had a world civilization taken "itself and its own perfectness with such disconcerting seriousness," nor tendered such lethal boasting, as the modern white world (Du Bois, 1971, 490). But those who suffered that world's aggrandizement were not deceived. "We looked at [the white man] clearly," Du Bois continues, "with world-old eyes, and saw simply a human being, weak and pitiable and cruel, even as we are and were" (Du Bois, 1971, 490). The demigods had feet of clay.

This theory of better, of white as preeminent right, however, did not only break out into bloodlust and launch world war. It worked itself under the skin, into the "warp and woof of daily thought"—with a thoroughness rarely realized (Du Bois, 1971, 496). Du Bois names it an "educated ignorance" that nonetheless pays and pays well. Rubber and ivory; cotton and coffee; gold and copper and bananas and all manner of product from the hell-pits of dark labor, which is given a pittance and early death as its own pay! Nations laughed, cities danced, art and science flourished, while "the groans that helped to nourish this civilization fell on deaf ears" (Du Bois, 1971, 492). European greatness, Du Bois cries, is not a greatness born from itself, but from "the foundations" and the past. Contrary to its own consciousness, it has raised itself upon the shoulders of its others: "the iron of ancient, black Africa, the religions and empire-building of yellow Asia, the art and science of the 'dago' Mediterranean shore, east, south, and west, as well as north" (Du Bois, 1971, 493).

But though "not-knowing," this "ignorance" is yet full of nuance. Color is shaded up and down its scale. Where dark yields descending difference from yellow to brown to black, from Asia through Amazonia to Africa, white yields right from south to north and east to west, from the Balkans and the beaches of Spain to Britain and the blond Dutch. The colonies were a solution to Europe's own internal conundrum: as white working classes climbed their own labor and education and political organization into a begrudged share of daylight with Anglo-French rule, darker labor became the new answer to the lost profits.

The imperial feast that quickly convened soon enough contracted envy in the wings. Germany and the Hungarians, Austria and Russia refused their own exclusion from the dark flesh-meal, and the mongrel blood-battle began. Whiteness had to admit new denizens. But the inner condition remained transparent to dark eyes. Du Bois's description (of World War I) exactly designates what awaits theological elaboration:

> As we saw the dead dimly through rifts of battle-smoke and heard
> faintly the cursings and accusations of blood brothers, we darker men
> said: This is not Europe gone mad; this is not aberration nor insanity;
> this *is* Europe; this seeming Terrible is the real soul of white culture—
> back of all culture—stripped and visible today. (Du Bois, 1971, 493)

And America in the dock admits of the same disposition. The "despising and robbing" of color that sits enthroned across the Atlantic also reigns here, says Du Bois (Du Bois, 1971, 499). It preys wherever white assumes might. Under its delusions of grandeur, however, a thunder sounds wide as the seven oceans, a storm of size and fury that will make "world" war seem tribal play. Darkness grows and whiteness, soon or late, will have its day of retribution (Du Bois, 1971, 501).

The Script Renamed: Whiteness in the Black Eye

Du Bois's lyrical lambaste offers an early profile from the century of color, a first widely published portrait of the reigning dissimulation in this land of light on the part of a dark scholarly voice. Where the nineteenth century had organized its white wealth-taking on the basis of bodies shackled in iron, the twentieth century needed a shackle of a different kind of mettle. Color-coding provided a continuation of the exploitation of labor under a much subtler regime than the chain and the whip. Du Bois proved prophetic in naming the line of color the curse of the day (as he did in the preface to the Jubilee edition of his first book; 1961, xiv). We have exited the century perhaps wiser to the wiles of mere perception, but no less enthralled, no less constrained and exercised. "Racialization" today has woven a garment with gender and economic rationalization that organizes a globe, shrouds sight in multihued and shifting lights of a projected "negritude," but continues to confer exemption from the scheme on white skin. Du Bois's reflections tracked an early exegesis: an "outing" of the invisible position, offered from inside the dark veil that has only in the last decade begun to find answer from Euro-American scholars with the emergence of whiteness studies in the 1990s (as we shall see in chapter 6). Black "writing back" in the 1920s remains prescient and pedagogical.

The picture is not pretty. Culling from the clairvoyance of the lyricism just reviewed, we find a litany of properties. Property itself is perhaps the foremost meaning of white "right." Du Bois is not amused—though he begins by noting the bathos of the belligerence. White styles of strutting, at first blush, *do* appear ludicrous. But the ensuing characterization of this "white religion of ownership" exposes the darkness of the deliverance promised: the large lie of white monopoly on accomplishment; the sadness of the ascribed self-importance—until the unconscious egoism is called out by black bitterness; then the eruption of hate and vituperation beyond any understandable calculus. Du Bois is pushed to grand metaphor and natural spectacle: white faces exhibiting a "writhing of hatred" vast by its very vagueness, "billowing" upward from the green world-waters like a great mass, wanting blood, praying pestilence on the humble, snarling in tigerish rage. It is a hatred taking fierce joy in its visitations of torture on black bodies, a fantasy that—strangely, tellingly—Du Bois finds to be a pitiable condition of "imprisonment" and "misery." The writing style is not incidental to the witness.

Even when turning to analysis, Du Bois retains the poesy as necessary to adequately convey the passion provoked. Whiteness is also a moral tenet, a boastful penchant, an aspiration to demigod-ery precisely in its demagoguery. Its church is business—a brokering of well-being on a wider scale, Du Bois is able to admit, but harboring in its brokerage such a widening of robbery, that earlier evils appear by compare as "saner" and more honest by far. And white business' "chiefest" business, in Du Boisian insight, is fighting—different from the past not only in brutality and scale, but especially in the callous ignorance of its beneficiaries. Underneath the bombast and blade, Du Bois finds cold calculus—the theory of white-over-black that licenses colony and empire alike, vaunting French and Brit over an entire world with impunity, consigning darker skin to mere laboring status—and early disappearance—as devil-child. The words pile up: "greatness" and "goodness," "efficiency" and "honesty," and "honorable-ness" to a fault, for whiteness; "bad," "blundering," "cheating," and foul for yellow; an awful taste for brown; the devil himself for black (Du Bois, 1971, 496). It is a scheme ever changing, but ever accumulating its bank account. Like locusts whiteness descends upon the lands of color, like kings itching after their El Dorados, sniffing and feasting like hounds, quivering on leash, or crouching snarling over their bones of prey. In self-honed theory, however, the dog is deity and the doctrine a "divine right of white people to steal" (Du Bois, 1920, 498).

And it is tempting to react. "William," we would say, "you exaggerate, we know better now than to pontificate in such piled up aphorisms, such extravagance of conceit, such metaphor! We are postmodern, ironic in the extreme, not given to purple prose or grand scheme, chary of mega-themes and meta-dreams of nightmares. You go too far!" But such a claim is itself at the heart of white meaning. Du Bois's language throws down a gauntlet exactly in throwing up a scream. Just what is the right word for this white world of proliferating violence? Inside the streams of teaming metaphor is a shrewd mental screening of the landscape: black is less than utter night; white is more than simple light. Du Bois' early offering here nuances the operation of race on both sides of the line of despair. Japan may be "yellow" in everyday congress, but when it dares snarl back, it begins to be tracked as black; Slav may be "dago"-tan from a position in the vanguard of Anglo-whiteness, but when the choice of neighbor is that or Afro-dark, white expands its ambit (without conceding the topmost rung). And Du Bois will not assent to an essential difference between the two terms of greatest distance; white weakness is recognizable because of his own; white greed is only a more virulent species of a common disease; white dissimulation is historically new only in its choice of "color" as the new rhetorics of rule. The uniqueness of the monstrosity is merely its size and extent—a world-system now, provoking sooner or later, a world-reaction.

But it is Du Bois's ascription of the word "Terrible" to this Promethean wantonness that names the condition most needing elaboration in white theology. It is the refusal to submit to tame terminology that demands scholarly redress. Du Bois has divined a skeleton: whiteness is a claim to property flamed by hatred; it is equally a claim to right underwritten by the darkness of its own active ignorance. These will frame the explorations that follow here. But it is the passion that Du Bois refuses to refuse, the terribleness, the sheer excess that his language invokes that sets the necessary tone. White self-consciousness can come to itself, I will claim, only by way of the refrains of delirium. It is not finally a form of rationality, though it masquerades in the daylight as the ultimate realization of enlightenment. It is rather dark and terror-driven, an underworld figment, an awfulness unfaced, thrown like a grenade of unconscious projection, onto an astonished world of plunder, whose rent bodies and regulated souls have yet to be fully recognized.

Unscripting: Looking into the Eye that Looks Back

Du Bois's writing serves as a touchstone for our explorations of race in outlining a phenomenological double-take on the state of the two souls he investigates. African American experience of "being made black" is apostrophized as a single body made to harbor a multiplicity of meaning, a doubling of awareness, a hypertrophy of consciousness, seeing itself from beyond itself. African American experience of the source of this incursion of color, this force that paints with such broad strokes of opprobrium, however, is an experience of terror, of a bared tooth of rapacity, an appetite that knows no surfeit. This is a mouth that opens the eye of its prey, a hunger that occasions—necessitates!— vision. Jonah sits inside the whale now, full of comprehension, spitting prophecy. Du Bois's writing itself teaches. On the one hand, as we shall see, his way of describing his double-consciousness strangely parries the initial thrust of the posited meaning of blackness not by resistance, but by a measure of "judo." Du Bois will neither block nor return the blow in kind, but absorb the energy and craft an alterity that looks back without labeling. On the other hand, his description of the white will-to-power pronounces its pox in images of animal lust or "divine" self-delusion. Whiteness scours the world like a hungry mongrel while parading in the mirror as more-than-human. Both strategies of representation also augur the necessary work in undoing white supremacy from within. Here, in a preliminary way, we could say, possession meets soul-loss. Blackness as "more-than-one"-ness in the same body finds its correlate in whiteness as "less-than-at-home" in its own head. And here, blackness is the first mirror of white self-recovery—not its savior, not its surrogate or solace, but its pedagogue. The work enjoined on the white side of the encounter is one of recovery of consciousness as multiplicity and of the body as human.

We can even begin an outline of discernment. The profile piqued by Du Bois's hand—later to be made explicit and explicitly confronted by Malcolm X and the Nation of Islam—is white supremacy as "Principality" in the Pauline sense (Eph. 6:12). Much work has been done in recent years to recover the biblical language of power—of forces large and labyrinthine, insinuating themselves in governments, in courts, in ethnic identities and imperial economies and sexual certainties, more than the sum of their constituents, an ethos of insidious influence, conforming thought and body alike to their constraining imperatives. Whether we sit easy or anxious to the notion of the demonic, "spirit" in the sense of a pervading climate of culture, proves eloquent to think with. The inside whisper of whiteness, the breath of supremacy on the cheek of identity, is a spirit blanketing an entire world now like an invisible mist. For far too many, it is in fact a death shroud. Black church practices, however—as indeed black musics and ebony theatrics and sepia politics and high yellow polemics and red-bone bombastics in a book, on a base guitar, hanging like time itself above a basket—have worked up such a down-pressing "heaviness of manners" into a communal vitality that bears witness. (John Coltrane and Bessie Smith, Paul Robeson and Lena Horne, Martin King and Fannie Lou Hamer, Adam Clayton Powell and Angela Davis, Toni Morrison and Ralph Ellison and Jimi Hendrix and Michael Jordan, all reflect the cultural eloquence of the costly effort to survive "in spite of").White supremacy can be made to yield "gettin'-down-ugly" beauty in spite of itself. It has also been made to yield Du Bois's writing.

Deconstructing the Genesis of Racialization

Reduplicating the Difference: Representing Race in Writing

Du Bois offers his famous formula out of an experience of having been made to feel different as a schoolboy. In probing his description of that experience as offering a particularly clear representation of the phenomenon of racialization, it is critical to pay especially close attention to what it indicates about the relationship *between* black and white. The moment when the "veil fell,"—the interruption of the exchange of greeting cards between the young Du Bois and the little girl—is represented by Du Bois as a very fraught moment of experience, when much of significance happens very rapidly. Illuminated by deconstructive analysis, the moment discloses a formal structure of interdependence in the same instant that it displays a particular experience of difference. The little girl's peremptory glance effects a "twoness" between herself and Du Bois that simultaneously separates him from her *and paradoxically joins her to him*. It effects a disjuncture that is simultaneously and necessarily a conjuncture. As we have already seen, his account is very careful—an autobiographical report of racism's advent in his own life that refuses its tactics in the very moment of representing its effects.

What could be called the original rupture of race for Du Bois—its moment of painful genesis in his consciousness—is not initially described in terms of color. Its originating gesture is narrated simply as a "peremptory halt" in an exchange—a "unilateral stop" initiated by a gaze. Precisely at the point where Du Bois's readers might expect to find race carefully articulated it remains unexpressed. In a certain sense, he writes its absence (Chandler, 1996a, 257). As pointed out above, it is thus a display that moves the production of color outside of the text itself and into the eye of the reader. The girl's cold eye opens up a space of blank difference between herself and Du Bois, but it is we who fill in the blank with a meaning of race. We as readers can presume the meaning because we know that eye from within. We also have opened up blank spaces and filled in between lines.

The Originating Moment of the Double: A Difference Between Two

But we are particularly interested here in the interior dialectic of this racialized exchange. Chandler's Derridian exploration of Du Bois's text is especially helpful at this point in slowing down the transaction we are investigating and examining its symbolic economy in detail.

For Chandler, race-making operates as a concrete historical instance—and deformation—of a more general phenomenon (Chandler, 1996a, 235–238). The production of African American subjectivity raises the question of the status of subjectivity itself. It begs to be situated theoretically in the context of the more transcendental circulation of identity and difference that constitutes the very possibility of human meaning-making in the first place (Chandler, 1996a, 243, 246). In a theory of signifying like that of Jacques Derrida, "difference" functions constitutively (Derrida, 1978, 279). It is part of the ceaseless irruption of experience and underwrites every attempt by human beings to identify that experience. "Identification" takes place by embracing one thing "as" another: I say "this is me": this body, this word, this posture, this style, this history, this feeling, this memory, is what I invest with the subjective identification "I." Difference (Derrida's *differance*) here hovers on the border between the "this" and the "me." It separates as it joins, and it defers the actual moment of identity across the space and time opened up between the two terms by the copula "is." Likewise with the negative coordinate of identification: the statement "this is not me" also operates by way of a difference-making that now joins what it separates. And this simultaneous joining/separating itself refuses specification in either example.

Difference cannot be pinned down and itself identified except by way of another differential operation of signifying: "the difference we are talking about is *this*"; or "the difference in question here is *not that*." It is the invisible gap between a word and what the word points to (in Sausserian linguistic terms, between the signifier and the signified). And what any word points to can only ever be "known" as another word. Things emerge for us—out of the buzz and flux around us—as distinctly "this" or "that" only by way of language, by means of the signs that render them distinct and different from everything else. Language always signifies across the gap of a difference that it both brings into being and hides. In consequence, identification never quite arrives at itself; it never quite manages to exclude difference from the sameness it seeks to bring to coherence. Rather, the attempt to identify something is driven to signify its sameness ("I" am this "body"; "this body" is "pink-skinned"; "pink" is a "color"; etc.), and what that sameness is different from ("I" am not a "tree"; not a "tee"; not a "three"; etc.), again and again. Difference remains "inside" the act of identification—a kind of interminable "ghost" that cannot quite be either named or eliminated—that is the very structure of signifying itself. In a very real sense, difference *is* language. And it promotes a constant circulation.

But in the moment of racial exclusion, a different kind of "differencing" is given license. Du Bois characterizes the moment as directed by a particular kind of envisionment, a peremptory glance. It is experienced as the violence of a "halt," the interruption of an exchange of greeting cards. At a deeper level, we could say that what is interrupted, or at least deformed or constrained, is the exchange of meaning, of identity and difference, that characterizes and constitutes the give-and-take of human relationship at every level (Chandler, 1996a, 237–238, 248).

The peremptoriness Du Bois describes then shows itself as a one-sided attempt on the part of the girl to halt difference "there" (on Du Bois) and thus identity "here" (in herself). Du Bois feels the effects of the violence instantly: it dawns upon him "with a certain suddenness" that "I was *different* from the others; or *like* mayhap . . . but shut out from their world by a vast veil" (Du Bois, 1961, 16; emphasis added). What had been a taken-for-granted flow between self and other, an ongoing interweaving of difference and likeness, is suddenly jolted. Du Bois is aware of "difference" now sticking to him, even as, in the same instant, he continues to be able to recognize likeness. What had been a flow is now an oscillation, an irresolvable confusion. "Am I different? But I am still like. But I *am* different; I must be, something has changed, a veil has fallen, a world has closed" (we could perhaps ad lib for Du Bois). What has happened is a motivated and one-sided stabilization of the uncertainty (of difference) shuttling between identity and otherness (Chandler, 1996a, 253). Difference itself is "incarcerated," we could say, in a lightning-flash presupposition. The little girl's gaze has interrupted the plasticity of time and space between herself and Du Bois with a presupposition about bodies.

The Time and Space of the Double: Originality and Purity

This presupposition of a "fixed" difference—a difference that is and has been stable, that has an identifiable history as "the difference that is race"—effects a violence in time and space (Chandler, 1996a, 251). The girl's refusing gaze is not just a simple "no," an *ad hoc* refusal that would let some other exchange be jointly negotiated. It is a "no, *because*"—a "no" that *knows* something specific. The human eye is capable of quite qualitative looks. Here, the content of the look is indeed a history, an ideology, a long habit of seeing. But what that eye knows, the very history presupposed in the girl's refusal, is itself brought into being *only in that very gesture* (Chandler, 1996a, 242, 271, ft. nt. 24). It is a "repetition" that renews (1996a, 246–247; 1996b, 80–81, 89, ft. nt. 3). The content of the little girl's look has no being in the time and space between herself and the little boy who stands before her with his card outstretched to her until it materializes in her eye and is made to adhere to his body as a definitive mark of identity, here and now, a new form of "I" solidified for him by her in the present.

And what is remarkable—what is indeed a mark of the very humanness of that moment in the way Du Bois presents it—is that the boy knows it, instantly. Even as a child, he knows the quality, the depth, of what has just transpired. He knows that it rearranges time and space around him. At some level, he experiences the fact that it "worlds" (Du Bois, 1961, 16; Chandler, 1996b, 80). And he recognizes that it does so "stereoscopically," world-ing two worlds at the border of the difference presupposed about the one. It is a profound gesture, even miraculous in a certain sense (which is why I turned to myth in my description of it in the early part of this chapter): a line of sight giving birth to two spheres, whole and original, from out of its own unidimensionality. It is a difference that is made to leap out of language—condensing time to a point, displacing space into an enclosure—and into the shocked substantiality of a body. In the blink of an eye, black difference and white sameness are brought into being "whole" as both preexisting the encounter and as autonomously homogeneous in their makeup (even if, in fact, they were only really created afresh, in the moment of actual encounter, as functions of one another).

The Structure of the Double: A Difference that Joins What it Divides

But something else happens as well. In the recognition Chandler's analysis opens, the intervention effected by the girl's refusal necessarily operates on and indeed signifies something in common between herself and Du Bois. It does accomplish a separation, by interruption, of two terms (two "worlds," in Du Bois's text). But those two are, in that same act, joined. They are paradoxically gathered together in the very force of their separation as ontologically intra-related (Chandler, 1996a, 261–262). The metaphysics is accomplished, as always, in discourse.

"Difference" here is not, cannot be, successfully sequestered in blackness alone, but continues to work its semiosis in-between black and white. Language, indeed, the very concept of race, is present in the girl's eye constituting and specifying its intention to divide.[5] She would not know either *to* refuse, or *what* to refuse, apart from the words that identify what she is differentiating. Du Bois does not tell us what those exact words are; we are left to fill in the blanks out of our own presuppositions and experience. Clearly however, whatever they are, they create a meaning of race, they signal a visual import generalizable as "black difference." But that very difference itself remains merely an effect of language. Its ability to signify, to convey meaning, depends upon its structural interlinkage with other terms it either displaces or eclipses. By itself blackness means nothing, it is blank; its specific significance accrues to it only out of the chain of other terms it subtly mobilizes or excludes. And as such, it is a difference that, like any other, functions *between* things—between terms, words, signifiers; indeed, between the girl and Du Bois.

Thus, in this reading of racialization, the play of difference intrinsic to all signifying, all meaning, is deformed and distended by an impossible intention (an intention that, in some sense, *is* whiteness). In the very attempt to halt the flow of difference that flow is made to overburden a single point or surface of significance. It overdetermines a single sign (blackness) with the pressure of a ceaseless proliferation. The attempted interdiction itself thus only sharpens the forcefulness, condensing difference into violence. Or perhaps we should more accurately say, the interdiction itself *is* the fundamental violence. The force-able hardening of discursive exchange into an ontological hierarchy—the impossible double realm of "absolute" difference and its concomitant of "pure" sameness—is the structural mechanism out of which the violence of racism is generated. (And indeed, this characterization also points the way toward one of the kinds of action necessary to begin overcoming racial violence—the necessity for white people to learn how to cease constructing the "halt," to reenter the flow of signifying, to reestablish rhythmic cadence in discourse, in bodily gesture, in perception itself, in the encounter with people of color).

Filling in the Blank: What is Looking Through?

In the eye of Du Bois's young friend, an entire structure of race has been reproduced, in a single glance. Words have appeared, fluttered, landed hard, words without a tongue, painting his body in meanings it will take an adult to articulate. He has been shrouded in density, incarcerated in a color, locked in a world-prison, all in a moment of sight. The seeing, in the first instance, was not his own. He only saw that he was seen, like catching sight of oneself in a mirror, in a moon-shadow, in crosshairs. He now knows even without knowing what he knows. Later he will write, giving tongue to the tautology, the tyranny that takes over his body. It is worth asking what fell in that moment of veiling. It is a strange moment, apocalyptic in portent, unveiling a secret like all good apocalypse—but here precisely in covering over what it renders naked. But the nakedness knows its display. The "scene" is a boy, now "black." The seer a girl, not yet knowing that she now knows herself as white. Between them a history, charged with numinosity, with predatory danger, a dark pollution demanding interdiction. The eye has been made prophylactic, warding off the seed. Would she be soiled? Assaulted even in prepubescent exchange? Whence this sharp sheathing of fear, this contraction of a hand, refusing a card, inside an eye?

We can only imagine, but it is an imagination well worn with fact. Relations between sexes across races have a sharp history in modernity, and in American modernity especially. Here, a girl encountering a boy in planned exchange. Undoubtedly parental anxiety has intervened, whether immediate and direct or long-standing and in the past. The history of white fear—perhaps the deepest meaning of whiteness itself—is clear: a male terror of male trauma, should a level playing field ever intrude between black and white. Ku Klux Klan grand wizards had regularly intoned the refrain: the chain must be maintained, by lynch-rope if not law, or a "bastard, mongrel race" is in the offing. But what is thus confessed? Either projection or preference. Plantation life had been systematically structured in white male rape of black female slave in the presence of the black male, as a "disciplinary" dismemberment of the black family—extra insurance against revolt! It was a form of systematic psychological warfare that returned like a form of haunt in the white male night, threatening post–Civil War revenge. It was almost unthinkable—psychologically—that the same would not be visited back on the perpetrator. Here the black male was made in white male mold in the white male mind: a blackness that is white! Or even worse, the white woman herself might well prefer the powers of black male prowess—the richness of tongue, the litheness of limb, the sharp intelligence born of a lifetime of intelligence-gathering—forged in the fires of duress, under the whip, in the ghetto, against impossible odds, creating with all the intensity born of desperation. If the playing field were ever really leveled . . .

The girl refused Du Bois's card, peremptorily, with a glance. And Du Bois, confused, knew his body was suddenly an entire history, bloated with significance, billowing out to the very edges of meaning with its blood, its beating heart, its boyish, growing strength. In her eye: her father's fear, her mother's mirrored ambivalence—repulsion? desire? both?—a family unconsciousness of world-historical devilishness. Was there a Principality in there? In any case, the boy must be castrated. So much content in such a small moment.

Du Bois's Doubleness as a Tactics of Resistance: The Question of Agency

Du Bois's strategy in dealing with the duress is brought out by noting how carefully he renders account of the experience. What appears immediately conspicuous is the absence of any attempt on Du Bois's part to negate the little girl's meaning (indeed Du Bois will immediately generalize: "I had thereafter no desire to tear down the veil"). Instead, the source of that meaning—the girl's glance—remains unqualified. And the significance it intends remains initially unspecified. Only subsequently are the dotted lines filled in.

Du Bois himself—separate from the girl—takes up the bifurcation that her glance inflicts in double mode: it cuts *between* them *and in* himself. He articulates its effect as one of "separating worlds" and creating "double-consciousness." In so doing, he gives her gaze—itself habituated in a language and a history—its first words. He speaks her meaning. But he does so only by way of reflection, as a reflection on, and of, what he sees in her eye. Remarkably, then, Du Bois's account of his first experience of racialization under white gaze reiterates neither whiteness nor blackness. And his "doubling" in this little girl's eye (as "being different") includes his own agency. He is no mere victim, here, not a purely passive recipient of her meaning. But it is a strange moment, this moment of estrangement. Its meaning is indeed negotiated by Du Bois, but not as entirely free or "undetermined."[6] He displays himself narratively as neither entirely passive automaton nor active agent. Rather, in Chandler's terms, he shows himself "solicited" (Chandler, 1996b, 88, ft. nt. 1). He indeed writes the (racializing) glance as essentially irresistible ("it dawned upon me with a certain suddenness that I was different from the others;"). But a scant semi-colon later, he also writes himself under the (partial) liberation (and damnation) of an irresolvable uncertainty ("or like, mayhap, in heart and life and longing, but shut out...."). He continues to struggle inside of the difference he has been "marked" with as if the marking is not entirely definitive.

The narrative thus traces a troubled and interdependent production of the "mark" of race—without naming it explicitly—that simultaneously leaves the place of the racializing gaze itself unmarked. Du Bois does not name the girl "white" in his text. The latter remains unspoken for, a (generous) textual refusal to foreclose what the girl would say about herself (Chandler, again in oral discussion). In Du Bois's text, her subject position is not reinforced by any attempt on his part to either delineate or oppose the force of her own power of opposition. That force (of violent opposition) is immediately inscribed in his writing in only one of its poles: it crystallizes an experience of racialization as duplication, of blackness as doubleness. But precisely in refusing to inscribe the girl "in kind" as definitively (different from himself as) white, Du Bois the narrator marks himself as more than her mere opposite as black. He rather creates a blank space in his text that "speaks" without a signifier. In it, difference itself is allowed to resurface as an un-subdued possibility. The little girl is left neither the same nor different, but rather semiotic—a sign, still awaiting its significance. Du Bois, in this reading, both recuperates and resists his own racialization in a manner that liberates difference as more than mere opposition.

Analyzing the Discourse of Racialization

Other voices of blackness have figured similar moments of initiation in equally intense descriptions. Frantz Fanon describes a moment of being accosted on a Paris street by a young boy's cry of terror to his mother as an experience of dismemberment, his body sliced in three, splattered with blood, stretched across history, bulging with ancestors. Patricia Williams finds it erupting in a split-second encounter through a suddenly closed shop window, white face looking back, hard as a glass veil. Poet Lamont Stepto speaks of Vesuvius in the head, Central American wars in the mind, Hiroshima-burned outline of human on blasted boulder as new self-cipher, "see[ing] . . . like you never [will again] until meeting God almighty" face-to-face. Blackness, we could say, in these reports, *is* consciousness. Du Bois makes us aware that it is not so much consciousness of some particular thing as consciousness *of* consciousness—pure volatility, flame-without-air, comprehending parallel universes in a single blink. The space of such a moment is null, its time a galactic horizon, the body alive like death in every cell. Language comes later. Race is first of all absent its colors, or perhaps better said, a burst of all colors in the black of one. It is a blank. It is rupture.

But the blank begs filling, like a vacuum, sucking in wind. Blackness will emerge in response, as we shall see in chapter 5, as in part, the proliferation of language, the desperation/exhilaration of signifying, ricocheting across the infinities opened between the eye that sees and the eye that is seen, making prison into vision. This also will be consciousness, unhinged from fixtures of words, exploring like a Mars probe. Indeed. Exploring war with words, warring against the "burning whiteness," as Fanon says.

Du Bois helps us see that race is a blank spot, hosting war. The war has taken ever-changing form over the recent American centuries, giving rise to entire vocabularies of shadow. Du Bois himself was instrumental in forcing a shift, at century's turn, from the n-word of white coinage, to the more formal "Negro," capitalized like a tuxedo-ed body. New possibilities of meaning opened—a blackness of Ph.D., of tax-paying citizenry, of dignity. In no time, however, old associations were loaded into the new sign in dominant discourse, "Negro" once again meaning "n-gra," itself working a lightning- stroke of hot neon flashes like "ignorant," "lazy," "hazy-in-the-head," "shiftless." The categories incarcerated a labor force in cheapness . . . and large white returns. Later effort insisted on "colored" in good NAACP style and just as quickly succumbed to the old infections. Later still would come "African-American"—hyphenated, not hyphenated, upgraded to include middle-class-ness, upbraided for the same, dread-locked and dashiki-ed and degreed, but still not freed of all the old history. At each turn, as white experience of black continued in new social space, new associations were loaded into the old litany. When southern migration to northern city occasioned new survival struggles and losses, and new skills and thrills of "getting' over," the term-of-the-day began to harbor new stereotypes of old hypes like "street-dealer," "razor-wielder," "pimp," and "thief." The 1980s and 1990s would load in "gold-digger," "gang-banger," "welfare queen," and "crack-phene." And of course, the stereotype just as easily flipped backside up and conjured qualities as positive as a child at play: dark skin could also mean "ever-quipping," "full of laughs and rhythm," and "fast-stepping," or "quick-cutting," "hard-hitting," and "slam-dunking," or "well-hung," "wild-strung," "exotic," and "erotic" as a tropical drink—depending upon white fantasy of black proclivity in the club, on the court, or cavorting on the dance floor. But even these distorted. Neither individual effort nor individual deviation found recognition in the imagined capacity.

And whiteness, too, harbored unexamined meaning in the fleeting rebound of all the projections. Here the assumed sense—common, but not any more earned—was a litany of stereotypic "American" nicety: "Protestant-ly working," "family-loving," "law-abiding," "freedom-championing," "fairly dealing," "tax-paying," "pleasant-acting," "wealth-deserving" . . . But, of course, ask someone darker hued . . .

Between the two terms of (supposedly) absolute opposition—and the metonymic associations and metaphoric exclusions they set in motion—opened an entire field of chromatic struggle. Other ethnic groups of color—immigrants arrived, arriving, or only anticipating—already knew the score. And recognized it was war. The history of social aspiration in the last century has witnessed all manner of jockeying for power and position. The scheme of representation now changes daily. Chicano morphs into Latina, divides into Cubano and Salvadoran, Puerto Rican and Chilean, Nicaraguanya and businessman from Peru. "Brown Power" organizes the differences into an opportune sameness in moments of political expediency (but sins "essentialist" in the process). Chinese American is not to be confused with Filipina or Thai, Malay or Burmese, or Indonesian-on-the-run. "Native" is Sioux or Cree, Apache or Zuni, Ojibwa or Onondaga, or the Crow. "Yellow peril" doggerel of the 1940s yields to fascination with Japanese quality circle of the 1970s is eclipsed by the images of east Indian prowess in Silicon Valley or Korean store profits in the ghetto in our day. But very little change has happened in direction. The great Unmoved Mover at the top of the pyramid of power attracting all comers to its gated reserves of the dollar (in boardroom and bedroom alike) remains the mystery of whiteness. And black has never ceased to anchor the lowest echelon, the position to climb away from.

The choice to play "black" back into the mix in the 1960s, however, marked a new moment in this straining assortment as already heralded in chapter 3. The African American community took control of the most common term of its denigration and loaded it with its own meaning of "soul." "Black" was "blood" was "beautiful." The war of the word—and its wordless pontifications—went public. And became the model for other "others" on the make—not so much in detail as in general tactic. Once a label has landed, and been wrestled into an alternative mode of meaning-made-and-on-parade, it ceases to be mere fiction, mere fantasy of a scheme of (dominant culture) control, and becomes rather a potential source of communal pride.

Behind all of this buzz, however, this massaging and maneuvering, this political struggle over transcripts hidden and visible—brokering livings, deciding housing, hoarding children in ghettos poor *and* wealthy, landing batons on backs, cuffing hands to bars, gathering dollars behind gates and deciding fates with policies and guns—is the blank. The commotion only exhibits the reality of the Du Boisian ploy: there is no real substance to racial significance, even if there are real effects of violence visited on the labels' targets. There are also real innovations on the side of the divide forced to create to survive the violence.

5

Black Performance

Leaving, then, the white world, I have stepped within the Veil, raising it that you may view faintly its deeper recesses,—the meaning of its religion, the passion of its human sorrow, and the struggle of its greater souls.

—W. E. B. Du Bois (The Souls of Black Folk, v.)

The Rodney King beating was one of the most graphic transcripts of the white production of "blackness" ever captured on camera. But what remains largely unknown in the country at large, and certainly un-comprehended, is the significance of the gestures King made with his body immediately before being so violently "interrogated" by wood and steel and leather.[1] According to the officer who first halted King, California Patrolwoman Melanie Singer, in her testimony against the LAPD officers, King was acting a little silly, laughing and doing a little dance when he got out of his inexpensive Hyundai (Gilmore, 29; Baker, 1993, 42). Patrolman Stacie Koon's defense attorney offered a slightly thicker description from Koon's deposition for the Simi Valley trial. He repeated Koon's testimony that King was "on something"; "I saw him look through me," and when Singer told King to take his hands away from his butt, "he shook it at her . . . he shook it at her" (Gilmore, 29). And in retrospect an unanswerable question emerges. How much of the blackness King was made to bear in his beating issued from an official impotence—the paradox of armed authority uncertain before an unintelligible performance? Seemingly, "white power" was face-to-face with something it could sense, but neither interpret nor tolerate.

Certainly the batons were intended to restore intelligibility. The subsequent violence of the police officers "clarified" King as the wrong-colored body making the wrong motion in the wrong physical and cultural space with the wrong timing—without any words said. (Even if, post-LA, that official violence was itself clarified as the real criminality in that scene of crime.) But subsequent replays of the video also went bail for words. As literary critic Houston Baker noted, the one thing we were never presented with were King's own words (Baker, 1993, 42). Even in stagings sympathetic to King, the video was assumed to be transparent. King himself, Baker emphasizes, was placed on a (media) pedestal like the slaves of former times, whose scarred backsides were displayed by white organizers at abolitionist rallies as mute testament to the savagery of the institution. We were offered the "victim-King" as surface cipher: the quintessential post-modern "blank" awaiting its expert information (Baker, 1993, 43). And thus analysis of the incident ever since has perhaps veered danger-ously close to that original policing function, trying to force the body of blackness to speak, as it were, "for itself," without bothering to listen to its own voice. A beating with words . . . .

The possibilities of such "speaking for" are notoriously endless. How can we reconstruct the event that begins with King emerging from his car dancing? Was it incipient "political resistance" and its rejoinder from the position of official power? Or was it something else, something at once more and less active, a negotiation of time, a jockeying for space, street heurism in the "kinesthetic disguise of black minstrelsy," "ventures Baker," . . . "contented darky" hoping to show white folks his "be happy"? (Baker, 1993, 47). What was it that was at work in the King body and its blue-uniformed "subduers"? How many faces of intentionality or determination leer out at us, its late readers? Simply a bit of incorrigible "black sass" put in its place by white seriousness? An "aesthetic displacement" of the sobriety of "arrest" (that itself stands as an instance of what Allen Feldman calls "the political art of individ-ualizing disorder"; Feldman, 109)? The leaking "on stage," in a moment of spontaneous boldness, of a hidden transcript, insubordinate "flash" before dominant authority . . . and the response of that author-ity, acting out its own covert script of superiority (Scott, 2–4)?

And what laughed? A Bakhtinian instinct to go grotesque in the eye of order? Youth? Drugs? Racial "character"? Some kind of sexist dis-comfort before "female" authority? Or was it a kinky-ness having its moment of play in a "scene of bondage"? (Perhaps "they" feared/hoped "he" was shaking it at them?) We don't and can't know for sure (perhaps not even King himself could say for certain). But surely the moment constituted a hermeneutically dense event, a loaded scene, a charged conjuncture of circumstance and gesture, bodies and backup, badges and banter, authorities and discontents, history and its inter- preters, "space" both in and out of "time."

In the cultural common sense, however, any questioning of the (untaped) King shake has been shut down by a more public production of politicized black dance. Damian Williams's prime-time-live jig of joy after sideswiping Reginald Dennis's head with a brick on April 29, 1992 is now the quintessential icon of LA-on-the-rise. Its titilation- power has been seared into public memory in countless media repro- ductions, likewise devoid of speech. Without question, that earlier "King silliness" now finds its possibilities of significance submerged in the stereotype. The "savage in the city" is already spoken for.

Setting the Black Stage

In a sense, the burden of this chapter is one of restoring voice to some of the bodies ventriloquized by whiteness in the public production of blackness in America. Part of the argument being constructed here is that white identity does not come into being or exist by itself, but emerges as the antithesis of the "mythic darkness" it projects onto populations of color. How those populations have survived their stig- mata, how they have displaced the opprobrium, resisted the violence and re-figured the mythology, is not merely fascinating, but feeds back into white cultural practice as well. Much of that feedback is necessar- ily surreptitious, playing stereotypic associations back into the domi- nant cultural expectations in ways that both use the mythology as "cover" and remake the values into "something else." In the process, black cultural creativity in particular in America, emerges as the ulti- mate (though not only) test case for white people seeking to under- stand what their whiteness means and how it functions.

Whose Body? Which Tongue?: Dark Roots and Deep Soils

Only a thoroughgoing appreciation of black transformations of the violence of racialization can begin to open up theoretical space for "thinking whiteness" *de profoundis* because only in that encounter are the deepest historical projections of white fear fully "outed" and figured, named, and faced. It is finally the encounter with *whiteness as terror* in the mirror of blackface that I will argue sets the necessary tone for all the work that must be done in exorcising supremacy, confessing history, and remaking community. What native peoples and the blood-bearing soils of the land pose as the cry of an indigenous Abel against the Euro-American fable of "civilization," what women silently and in bruised scream offer as rebuke of the violence that is "intimate" to family life, African American bodies—struggling against the categories and criminalities they have been made to bear—agitate to the surface of whiteness as a supposedly benign form of identity. It is the deep blue waters of the Atlantic Middle Passage that remain the ultimate baptismal pool for white people wanting to understand—and repent—what race has meant historically.

Du Bois in chapter 4 named white culture as Terrible, under its skin of civilized enlightenment, a Terribleness most visibly seen in world war, but nevertheless hard at work in daily life in counting its capital and coercing its labor around the world. Until that Terribleness is given a face and a name under white skin, efforts to remedy the rancor erupting in communities and countries of color across the globe will fall short of the change necessary. It is here that black arts and resistance, black church and creativity, offer example and pedagogy. They are figurations of a struggle with force that counsel courage and teach possibility and speak inside the body. The courage concerns any effort that would too quickly seek to fix "the problem" before its depth has been augured and its *pathos* entered. The possibility concerns the flexibility of life's vitality before it is colonized in a category (like race). The body is the court of final appeal, the real zone of change. This chapter proceeds under the triple demand to investigate deeply and imagine broadly and explore in the language of muscle-and bone-in-motion the resolve to recover greater wholeness. The work of historian of religions Charles Long and feminist critic bell hooks on ritual transformation of white terror into black vigor, of English Studies adepts Henry Louis Gates, Jr. and Kristin Hunter Lattany on everyday improvisation of white signification into black affirmation, and of cultural studies scholars Paul Gilroy and Tricia Rose on rhythmic syncopation of white order into black pleasure, will uncover some of the requisite tactics.

The choice for nontheological scholarship here parallels the claim that secular practices of race today operate as inchoate modes of theology—theological aspirations operating in Enlightenment "drag," that find their most instructive confrontation as "white" in black practices that could themselves be understood as "popular culture spiritualities." The "trinity" of transformative work with ritual practice, speech pattern, and rhythmic performance impels a theory of whiteness to deal with questions of "white cosmology" (the world or space of whiteness), "white speech" (the temporality of whiteness), and "white silence" (the body of whiteness) that are already operating in power before theology as an academic discipline utters its first word. These domains of black practice together mark out and contest the practical dimensions of a surreptitious white soteriology.

Black Culture in the Making

In the compendium of essays written between 1967 and 1983 and published in 1986 as *Significations: Signs, Symbols, and Images in the Interpretation of Religion,* Charles Long develops a broad project in which he reads the experience of colonial subjugation as itself quintessentially "religious" for those who have been made to undergo Western expansion and exploitation. Long is important for the effort here especially in raising issue with the adequacy of theology as a mode of discourse to deal with the religious movements of the oppressed that result from the attempts of indigenous people to re-create their reality after wholesale dismemberment under the Western onslaught. While not focusing primarily on categories like white and black, his characterization of Western domination and indigenous response nevertheless delineates an entire global situation of rupture that has yet to find a methodology of study adequate to its subject.

Theologies opaque (Long's general category for the "Black" and "Red" theologies of James Cone and Vine Deloria, respectively) emerge in Long's writing as signs and symptoms of the time. On the one hand, they have indeed specified the problem of the modern Western project of "enlightenment" as one not only of bad behavior and bad acts but also of bad faith and bad knowledge. Against the vision of God supposedly made transparent in Jesus crucified, they have (importantly!) reasserted the opacity of the suffering and oppression found there—and in all of the other crucifieds in history. It is an opacity that mainstream theology as discourse has been unable to figure in its tremulous and anguished depths. But on the other hand, they have not yet radically broken with theology as itself a structure of power and a norm of discourse that must be decoded in favor of "prepar[ing] a place and a time for the full expression of those who have suffered alterity and oppression" (Long, 1986, 195). It is construction of this other time and space—and reeducation of the kind of body that can inhabit and express such—that must focus attention theoretically. Long's own effort to augur the depths of the dilemma and name the historical creativity that has been aroused in counter-point finds perhaps its most precise expression in a 1975 essay entitled "The Oppressive Elements in Religion and the Religions of the Oppressed."

Long on Dread: Mysterium Tremendum et Fascinosum

In this essay, Long juxtaposes the writings of William James and Du Bois to specify what is missing in European and American attempts to formulate religious experience—an experience that Long argues has been analyzed, in empirical approaches like that of James, as well as the theological approach of someone like Ernst Troeltsch, by way of a surreptitious norm of Protestant individualism. Long's own counter-project has been one of insisting that it is indigenous myth, rather than European science or theology, that most adequately figures the human scale and significance of the trauma resulting from colonialism and its desperate aftermath—and that it is folk creativity rather than empiri-cism that has facilitated survival of the same. What is missing from the Eurocentric approaches becomes clear only in asking after the *negative* in religious experience. This is the element that indigenous communi-ties have been forcibly "baptized" into, as a form of ongoing historical initiation, that neither European theory nor theology can adequately disclose in scientific language. Long himself does not eschew Euro-theory, but raids and redeploys its terms in service of an inchoateness of experience and opacity of anguish that finally require ritual work for their adequate expression and transformation. But the experience of this negative element is not exclusive to communities crushed in the press of colonialism.

Long points out that both William James and his father Henry James, Sr., offer personal accounts within the corpus of their works of moments when each experienced utter terror—the sudden rupture of an entirely mundane moment by a seemingly unmediated premonition of horror that briefly shattered their sense of self-possession (Long, 1986, 160). But they do so in a manner that dissembles. Henry, Sr. writes of a sudden dismay that engulfed him while lingering alone at the dinner table one evening following a perfectly normal repast with his family, when,

> in a lightning flash as it were—"fear came upon me, and trembling, which made all my bones shake." To all appearance it was a perfectly insane and abject terror, without ostensible cause, and only to be accounted for, to my perplexed imagination, by some *damned shape*

squatting invisible to me within the precincts of the room, and raying out from his fetid personality influences fatal to life.[2]

William, in turn, writes of a similarly unremarkable evening in his own domestic quarters, when, for no apparent reason, in his dressing room,

> Suddenly there fell upon me . . . a horrible fear of my own existence. Simultaneously there arose in my mind the image of an epileptic patient whom I had seen in the asylum, a black-haired youth with greenish skin, entirely idiotic . . . This image and my fear entered into a species of combination with each other. *That shape am I,* I felt, potentially. (James, 157).

Both men—white and privileged we might note—confess experiences of dread. But the very terms of their accounting also confess something else. The neurological–biological and individualistic categories invoked in representing the experience elide any role American society and culture could have played in transmitting the experience (Long, 1986, 161). For Long, the radical contingency James seeks to honor in his work is not adequately probed in individual psychological formulations like the "once born" or "twice born" formulas James is particularly famous for. Only something like Rudolph Otto's valorization of an *a priori nonrationality* of experience begins to get at the hidden depths of the question. It is a depth that is finally a mode of the divine itself (Long, 1986, 162).

Otto's formulation of *daemonic* dread as one aspect of the great mystery that is simultaneously Terrible and Fascinating (the *mysterium tremendum et fascinosum*) affords Long a way of talking about the possibility of religious experience prior to its schematization in a category. This feeling of numinous horror, of creaturely belittlement and unworthiness, of the overwhelming aspect of majesty leading to absolute dependence and humility, constituted, for Otto, an oppressive sense of the divine present in all religious traditions. It gives rise to an "internal meaning of oppression in religion" (Long, 1986, 163). Reformulated more precisely, later on, by Gerardus van der Leeuw, it takes shape as a vague Somewhat that generates significant meaning *by forcing itself upon and opposing itself to* human beings as Something Other (Long, 1986, 163; van der Leeuw, 23). Its effect in religious consciousness is a profound sense of being merely a creature—the radical experience of contingency before the awesomeness of an inscrutable Creator.

Long on Language: Du Boisian Double-Consciousness and Demonic Dread

For Long, the basic concept becomes a pliable heuristic tool by means of which he can evoke the shadowed depths of American experience concealed in the usual telling of its story. W. E. B. Du Bois supplies his data. From the *Souls of Black Folk*, Long quotes the passage in which Du Bois recounts his first experience of the religious worship of an oppressed southern black community when he was a teacher at Fisk:

> It was out in the country, far from my foster home, on a dark Sunday night. The road wandered from our rambling log-house up the stormy bed of a creek, past wheat and corn until we could hear dimly across the fields a rhythmic cadence of song—soft, thrilling, powerful, that swelled and died sorrowfully in our ears. I was a country school-teacher then, fresh from the East, and had never seen a Southern Negro revival . . . And so most striking to me as I approached the village and the little plain church perched aloft, was the air of intense excitement that possessed that mass of black folk. *A sort of suppressed terror hung in the air and seemed to seize us,—a pythian madness, a demonic possession, that lent terrible reality to song and word. The black and massive form of the preacher swayed and quivered as the words crowded to his lips and flew at us in singular eloquence.* (Du Bois, 1961, 140–41; emphasis Long)

Long is especially interested in the language used by Du Bois here. He notes, "in some particulars it sounds very similar to the kind of dread described by the [Jameses]" (Long, 1986, 164). But Du Bois's description reflects a difference that becomes a source of critique of the Jamesian accounts. Whereas James situates religious experience in individual feelings and acts in solitude, Du Bois is pushed into reflective rumination of a historical kind. Against the abstract description of philosophy, the beauty and sorrow of Du Bois's community moves him into concrete musing about Africa and a kind of *primordium* that is given deep amplification in the experiential terrors of slavery:

> The Music of the Negro religion is that plaintive rhythmic melody, with its touching minor cadences, which, despite caricature and defilement, still remains the most original and beautiful expression of human life and longing yet born on American soil. Sprung from the African forests, where its counterpart can still be heard, it was adapted, changed, and intensified by the tragic soul-life of the slave, until, under the stress of law and whip, it became the one true expression of a people's sorrow, despair, and hope. (Du Bois, 1961, 141)

The point for Long, however, is not that the difference be cast as one between solitude and sociology, American individualism and African communalism. Rather, Long is concerned to exegete black experience of the otherness of the situation in America as an ambiguously negative experience of the divine itself (Long, 1986, 164). It is a religious experience of divine mystery and terror that cannot be reduced to any other category even if it can be schematized into other meanings or notions (such as "colonization," "exploitation," "sin," etc.).

Thus, by way of Otto, Long is able to comprehend slave and ex-slave conversion accounts in terms of an experience of the *mysterium tremendum* that "is never identified with the sociological situation or with the oppression of slavery itself" (Long, 1986, 164–65). What is experienced in such moments of creaturely diminution is simultaneously an experience of an "essential humanity *not* given by the slave system or the master" (Long, 1986, 165; emphasis added). As Long is at pains to point out—the numinous does not just overwhelm and "oppress" the slaves; it also relativizes the oppression of their oppressors.

However, Long is quick to add that historical structures of oppression do color the experience (of this divine oppression) and "create a screen." Du Bois's description of such in terms of being born with a veil and plunged into the peculiar sensation of double-consciousness emerges from this experience as "at least a critical statement." For Long, such a statement "points to the ambiguity of community and religious experience at the very moment of perception" (Long, 1986, 165). Du Boisian double-consciousness, for him, "conveys in a *sober manner* the overtones of the *same sense of demonic terror and excitement* that Du Bois describes when he visits the black church" (Long, 1986, 165; emphasis added). It points not just to black struggles with white expectations, but also to black apperceptions of a quintessentially religious "Otherness" that will not let them alone even in the midst of their human battles. It speaks of an embodied struggle that also indexes a spiritual wrestling. But according to Long, the experience in question is *not* simply unique to the black community. A similar dread or terror also lies behind the account offered by the Jameses.

For Long, the difference hinted at in the language used respectively by Du Bois and James is to be found in the radicality of the contingency associated with the experience. Long is led by Du Bois's account to ask pointedly of the Jameses, "Wherein lay their concrete memory and *primordium*?" (Long, 1986, 165). The individual psychological and mystical categories they employ are not innocent, but *conceal* a memory that, if rendered concrete and public, would face the white

American community at large with its own radically contingent state. Dread is not simply an individual existential crisis, but has social correlates and historical coordinates.

Long on Theory: Du Bois by Way of Hegel; Hegel by Way of Du Bois

Long deepens his analysis by framing the specific comparison of Du Bois and the Jameses in a more general discussion of religions of the oppressed in relationship to modern Western philosophy. The appearance of these new global religious phenomena in the nineteenth century—also variously called cargo cults or millenarian movements in scholarly treatments—cannot be dismissed as "simply perennial forms of religious life" (Long, 1986, 166). Although bearing some resemblance to the crisis cults of earlier movements in religions like Judaism, Christianity, and Buddhism, these new movements of the oppressed take their specificity from the nature of modernity. In relationship to modernity understood as "itself . . . a form of critique," they constitute the "critique of the critique."

The cultures hosting such new religions are mostly ones that have been "created for a second time" in the theoretical disciplines of the West (Long, 1986, 166). It is this second creation that gives rise to the Du Boisian formulation of double-consciousness. But the experience of this second creation is also the source of the critique of the West. For the cultures reworked in Western scientific discourse, objectifying and distancing categories like "civilization," "primitive," "race," and so on offered no intimacy of language within which people could re-fashion an identity after the trauma of contact. Western rational discourse was experienced by the colonized more as the "quixotic manipulation of a fascinating trickster" than as something they could identify with or find themselves within (Long, 1986, 167). The imposed meanings occasioned a spurious apprehension of the *tremendum*—an experience of overwhelming contingency that was actually only "the other" of the other culture and not of the divine majesty. At the same time, however, "within the veil" that mysteriously dropped down on these peoples, separating them from their colonizers, a response was crafted that was not just intellectual, but experiential and total. Movements like the Ghost Dance, the Vailala Madness, the Rastafarians, and the Black Power movement created their own meanings. "Radical contingency" found its adequate expression (only) in the fear and trembling of "strange ritual."

To clarify the particular efficacy of Du Bois as a figure of transition who creatively combines both Western intellectual prowess and the innovative action of these cults, Long drafts Hegel's master/slave dialectic into his argument. *Contra* Lockian notions of slavery as peripheral to society and history, Long notes, Hegel places the primal combat that enslavement dramatizes at the natal core of human consciousness. For Long, the struggle for identity in colonized and neocolonized societies is precisely this struggle. It is an attempt on the part of the oppressed to create not just a new consciousness for themselves, but "a new form of [human] consciousness [in general] and thus a new historical community" (Long, 1986, 168). What is of particular cogency for Long is Hegel's dialectic reversal in which the slaves shape their own consciousness, independent of the master's definition, by means of their *labor*. In Hegel's rendition, it is by immersing themselves in nature and work that slaves move beyond the terror that initially engulfs them (Hegel, 117–119). By transforming nature into *products,* they finally objectify their subjective consciousness of their situation. Masters, on the other hand, remain mere consumers of products (of their slaves) that have in no way transformed their condition into autonomy or objectified their consciousness in reality.

Long places this Hegelian understanding at the center of his comparison of Du Bois and the Jameses to make clear—in redeployed *Enlightenment* terms—the difference between an "Enlightenment" apprehension of dread and the expression of the same in an oppressed community. Long first points out that the mode of expression of the Jameses is largely "fantastic." "The objective element that is the *other* in their experience is vague, ill-defined, abstract" (Long, 1986, 169). It is rhetorically represented in psychological terms "unrelated to existence." In the words of Henry, Sr., it is a "damned shape squatting invisible"; in William's account, a "combination" of the remembered image of an epileptic patient and his own fear. In their representations, the Jameses are masters primarily of their own solipsism.

In Du Bois, on the other hand, Long notes, not fantasy, but community is to the fore. It is a matter of listening to the words and sounds of *each other,* and whatever is fantastic in the experience for Du Bois arises from the *production* of the ex-slave community he is standing among:

The black and massive form of the preacher swayed and quivered as the words . . . flew at us . . . the people moaned and fluttered . . . the gaunt-cheeked brown woman next to me suddenly leaped into the air and shrieked like a lost soul...round about came wail and groan and outcry, and a scene of human passion such as I had never conceived before. (Du Bois, 1961, 141)

What is fascinating in this scene—and no mere fantasy—is "the immediacy of experiencing outside of veil and double-consciousness" that Du Bois is afforded (Long, 1986, 169). For Long, it is an immediacy that must be traced back to "the autonomous creation of the slave community." It has reference to an experience of self-intimacy and communal solidarity *not* determined by the category "slave" (or in the post-slavery community, by any of slavery's successor categories [such as "black"] controlled by the dominant white culture). The autonomy Long is underscoring comes into play in connection with the (Hegelian) dialectic in which slaves come to independent self-consciousness by means of the created objects of their labor.

But Long is creatively adapting, not just repeating, Hegel in this account. The labor that afforded slaves a measure of self-consciousness was not just that of menial work in the fields, but even more importantly, the ritual work carried out in the hush arbors and bush meetings. The result of that work is a mode of self-conscious experience "outside the veil" that is now perpetuated in the religious reproductions of the black church. The products of that labor are the communal styles of preaching, singing, moaning, shouting, etc. by which the ex-slave community continues to transfigure the reality and memory of oppression into totemic power. Dread here is not so much discharged in private sweats and shudders (as with the Jameses) as redirected in communally valorized forms of creativity. For Long, black religious rituals and traditions are forms of practice *constituted* in a struggle to transform demonic violence into *daemonic* vitality.

The Long and Short of Modernity

Long's language—and the critique of Enlightenment-based empirical or theological language that it intimates—invokes a depth that is by definition difficult to define. Like Du Bois in chapter 4, he gropes for terms that measure up to the kind of complex convolutions of experience that do not readily submit to language—that are, in fact, distorted to the degree they are formulated in coherent terms. Du Bois ranted. Long raids. The figure here is not fascination, but terror, an intimation of God that overwhelms and shivers. The innovation is that the suffering marked by blackness (as well as other forms of otherness) is not merely a suffering of white oppression, but rather finally a suffering of "God." It is the experience of colonial and postcolonial oppression as that oppression is itself haunted by something even more Terrible: an apparition of the Ultimate Other that is somehow also present in the trauma, as the real Terror, the kind of Dread presence that will also allow white people to perish. This is a (non)face of God that the oppressed alone know.[3]

Dominant American culture—including dominant forms of Christianity and theology—are at this level mythically illiterate. Myth is the modality that deals with structuring impossible contradiction into "inhabitable" forms of narration. And ritual labor is the incubation chamber and hothouse living room of myth. European worldviews, in the colonial encounters in Africa, Asia, and the Americas, were not immediately ruptured in the way indigenous myths of origin were—and in consequence, modern Europe and America remain in some sense without a myth of modern origins. They misperceive themselves as merely in continuity with the Greeks of old or the Renaissance of late.[4]

Indigenous communities on the other hand, had to re-create themselves from a new starting point called modernity. The West came crashing in on them, violating not only bodies, but bearings, not only muscle, but mentality. Their entire orientation to the cosmos was blown apart, shredded, rendered inert. Reality became "opaque," incomprehensible. Survival dictated not only feverish exertion for the master culture materially, but furious imagination, behind the veil, to reconstruct a habitable "world." The result can hardly be represented in writing; its real modality is ritual concatenation, bodies writhing under possession by the West, roiling in the tensions of old identities pressed into new molds, of inscrutable powers suddenly made palpable and painful in the form of chains on ankles, whips on backs, books that talk, bullets that break bone, an experience of being "eaten" in one's labors by a tooth that never quite appears, but consumes the product somewhere else. In a word, indigenous ritual had to work out a meaning of Reality as Terror with the community alone as the requisite text. The result could be said to be possession cults that give human form to the opacity of Otto's *tremendum* and in so doing come to a knowledge of human "being" that is beyond words.

But Long's descriptions are telling. It is not that the oppressed alone die or fear dismemberment or insanity. It is rather that they alone are forced to face such without recourse to human prophylaxis. They cannot place other human bodies between themselves and the hardness of gravity. In one sense whiteness emerges in the colonial encounter as just such an operation of superiority, an attempt to use darker bodies as a denial structure, a medium between rocky soil and ready food, between hard labor and coveted leisure, between death and the living that inevitably lives toward such an end in the grave. The Jameses also know terror, but their narrative of reality, their resources for integrating the upwelling of dread, are beholden to science and data, civilization and decorum. They are reduced to moaning in secret as mere individuals rather than gathering their delirium into a communal emporium of the groan, a shared space of grappling with contingency and vulnerability in idioms adequate to the pain. It is this (!) that black church performance performs and that Du Bois, enfolded into their delirium, embraces as a "black depth" heretofore alien to his experience as educated and middle class. It is art, not science, and at heart, it is more human than merely sober worship.

But there is more to be said. The particularity of this terror, the historical composition of this "Principality," is composite, a ghosting haunting the peculiar rupture that is race. It is the business of myths of origin to "humanize" ruptures, to explain the inexplicable experience of duress and devilry that accompanies every culture's attempts to survive. Why there should be death, why the gods are remote, why miscarriage and misfortune, mistakes and earthquakes interrupt the struggle of culture to regularize existence, is made both tolerable and inscrutable by mythic recounting of some tiny infraction of the past— peeling the banana at the wrong time, stepping into the wrong puddle, eating the apple of the wrong tree—that memorializes the break of coherence. Such myths do not so much tame the terror as give it a body to live in inside the community.

But the body of whiteness has never had such a mythic venue to work out its fright. It dreads "dread" and refuses to lend it a body to speak through. Or rather it projects its dread onto the people it identifies as the living embodiments of dreadfulness and tries to "surrogate" the encounter through them. At this level, white identity appears as a kind of bodiless head, moving through the spaces of global society as a project of fascination and innocence, running after the latest fashion, coveting the most extreme titillation, building entire industries attempting to mute the terrifying whisper within through "entertainment." And of course, to the degree that other cultures have been swept up into Western desires for development and escape from death, the break between the human body and its ultimate destiny has not remained merely a symptom of whiteness. But the colonial and postcolonial formation of white identity structures remains racially normative for the historical intention I am here glossing as a project of "global denial" as we shall see in subsequent chapters. Suffice it to say at this point that one of the very meanings of whiteness as supremacy has been an attempt to escape the terrors of contingency by, in effect, *forcing* other populations to know that particular experience of creatureliness intimately. Dread and terror are not thereby rendered uniquely the experience of the oppressed, but the communal ability to apotheosize, liturgize, and humanize such in dramatic forms of creativity are.

In this sense, white people are beholden to communities of color for mythic sustenance in this age of the individual. There is a hunger for depth inside the suburban teenager listening to Charlie Parker or Tupac Shakur that cannot be explained away simply as fantasy and immaturity—though these latter do themselves say something crucial about the dominant culture in America. And this too, Long locks onto. Absent ritual congress with its origins in the modern moment of encounter with its others, white culture is reduced to an apperception of the Ultimate as only Fascinating, not tremendous. The result is a culture of triviality, a constant search for allure, for seduction. And a deep fear of fear that is, ironically, the triumph of fear itself. Otto's aphorism is illuminating. As Long has so ingeniously argued, the West only knows God as *fascinans;* God as *tremendum* awaits honest encounter, and uncompromising labor, in relationship with those who never, in modern times, have not known the Ultimate in that aspect.

Tactical Performances[5]

What Long has given to the academy is the possibility of recognizing that, like Du Bois' experience of the little girl's gaze in his childhood initiation into the rupture of race, European-based protocols of theory also embody a gaze that "halts." The remedy for that halt is not so much refutation by way of argumentation as embrace of a wholly different order of knowledge. Black consciousness is an awareness that has been forged in the crucible of constraint: the experience of social and psychic violation that historically had no public space for counter-expression other than the gathered community of bodies similarly violated by the rejecting eye and hand (or iron chain or steel bar or penis or rope or bullet). This is consciousness doubled, coiled in on itself like a Kundalini snake of energy awaiting awakening, consciousness made hyperconsciously aware, visionary even, that sees below the surface of white society into the opaque depths of the Terror harboring there that assumes divine-demonic shape. It is a Terror that cannot be exorcized by being merely analyzed, but can only be obliquely recognized and managed by way of myth, and the ritual body made aware of myth in its cultural core. It is precisely a *myth of origins* that Du Bois's account of his encounter offers, a story encoding an experience of "primordial" rupture that cannot be simply wished away. It is a myth of origins of blackness that is also the genesis of the whiteness that created such blackness in the first place.

Ongoing work with this Terror, however, has been forced on the black community by a white community structured to avoid such as much as possible. The question such "body-work" raises for the white academic is necessarily a question of the white body and its spaces of habituation, its motions of ritual, its codes of timing. It is these more subterranean codes of enculturation where myth lives in muscle and bone that are in fact "outed" by black cultural performance and brought to the surface of awareness for the consciousness cued to look for such. Subsequent chapters examine these "codes of whiteness" in their quotidian appearance in the social order. For now, however, it is essential to name them mythically as a mode of rupture and spiritually as a meaning of *Tremendum*. Everyday experience of encounter with black people opens the possibility that at any moment the encounter could plunge through the veneer of polite exchange or not-so-polite avoidance into this other realm of the terrifying. There is no way to undo white fear of otherness except by facing the fear inside the body. At stake is a question of passion, of energy currently locked up in structures of denial and control (both psychic and social, the white body going rigid in the elevator when a black male enters, the gated community hunkered down behind the "thin blue line" when the city erupts). At issue is the question of the "place" of such white work, the social space in which such fear might be explored and exorcized, named and converted to passionate involvement in building a different kind of world. Before engaging such work, however, it is important at least to outline other modalities of black transfigurations of this energy of supremacy, other ritual tactics reconfiguring white violence into a deepened capacity for experience both human and divine.

bell hooks and Black Looks

When we turn elsewhere for witness on this phenomenon, we find similar themes elaborated with distinctive leanings. African American cultural critic bell hooks also exegetes the glance of race and finds terror. In her book *Black Looks: Race and Representation*, she interrogates

the power of the human gaze as both captivity and contestation—
a power first discovered in childhood resistance to parental control, a
looking back at adults that, for hooks, quickly drew censure and punishment (hooks, 1992, 115–116). She eventually links that family
sanction to the tradition of southern white supremacy demanding that
black slaves and later, black "citizens," never look directly at white
people—lamenting that her own community had reproduced this tactic of domination in its own socializing process. But the lament also
led to revelation. hooks gradually discovered the politics of looks and
the capacity to translate viewing the (white) other into empowerment
of her black self by *continuously looking into her own experience of
looking* and making choices about when and how to gaze back at the
intention to dominate. By tracing, through postcolonial and black literature and film criticism, the possibility of a return gaze that did not
merely reproduce the stereotype in reverse, she was able to identify an
alternative "syntax of recognition"—black eye exchanging information, affirmation, and mutual contestation with black eye (such as
I identified of the women in my classroom in Chapter 1), absent words
that would dangerously disclose too much (hooks, 1992, 129). But
hooks was also able to identify the gaze-that-dominates. It was finally
male as well as white—an ideological eye constructing social desire in
the form of a sexual "object" that was ideally blond and pale and
exhibiting the attributes of Barbie. This was an eye of commerce and
surveillance that a "black" woman would never entirely be able to
satisfy.

When turning more directly to the question of how to represent the
subject of this gaze, the bearer of the look that dominates, hooks had
to learn to discern mere reverse stereotypes from more ingrained psychic states among her people (hooks, 1992, 169). She remembered her
own experience of having to walk across town to visit her grandparents, a walk that took her through white streets of hard stares and real
danger, a space of fear that she now experiences as inscribed in a
globe, facing similar scrutiny and uncertainty—and the possibility of
interrogation and even arrest—whenever crossing national boundaries
in airports (hooks, 1992, 170–171, 174). Whiteness for hooks and the
southern black folks she grew up among was connected with "the
mysterious, the strange, and the terrible," she says, a physiognomy not
readily known, appearing with terrifying suddenness in segregated
"black spaces" to sell products or Bibles . . . or wearing sheets
and bearing torches (hooks, 1992, 166). Even when friends or neighbors sought to assure some measure of recognition by means of

imitation—by "acting white"—suspicion and fear were in the forefront of concern. "White knowledge" was coveted not just for itself,
but also as a kind of amulet or mask that could be worn to ward off
the evil (hooks, 1992, 166).

Hooks's own antidote for such emerges from her encounter with
the witness of women like northern born Itabari Njeri, who writes of
traveling south to investigate the murder of her grandfather by white
youth who ran him down in a drag race. For Njeri, the journey became
a kind of ritualized revisitation of terror through reenactment: a pilgrimage out of forgetfulness and ignorance of (her own) history back
through the bitterness of loss to the scene of the crime to "face the
enemy." When she finally did come face-to-face with the white man
who was said to be the murderer, she saw her own fear in his eye—and
perhaps even the "mirror [image of] the look of the unsuspecting black
man whose death history does not name or record" (hooks, 1992,
172). Such travels enact an archaeology of memory, offers hooks, confronting the debris of time and the pain of loss with the kind of active
solicitation and work of reconstruction that alone are able to "force
the terror of history to loosen its grip" (hooks, 1992, 172). She links
such accounts to her own childhood and adult passages through the
many spaces of "terrifying whiteness," and broadens them both to
include the collective journeying of black people in the Middle Passage
and in mass migrations north during the two world wars, to assert,
with Michel Foucault and Jonathan Arac, that such work reclaims
memory as a site of resistance, transforming history from mere judgment on the past into a counter-memory that combats the present.

Alongside such ritual revisitations of the memory (and continuing
possibility) of terror, hooks also places a more subtle though no less
troubling recent experience, occasioned by supposed allies in the
struggle who—at a conference on antiracist activism—unwittingly
reproduced the hierarchy of white supremacy in protocols about who
was allowed to speak, who got to sit where, and whose body language
controlled the discourse (hooks, 1992, 176). Having labored lifelong
to uncover the profound connections between mundane experience
(e.g., a mere glance) and large-scale event (e.g., the Middle Passage),
hooks could not suddenly shut off the awareness—or the fear—just
because the protagonists claimed alliance. Indeed, such a profession
heightened the harrowing. If even *these* white folks remained blind,
what hope was there? When hooks publicly confessed her fear to her
white co-participants, the experience was laughed off—a response
only further reinforcing the feeling.

The same conference, however, afforded interaction with a black female co-sufferer of the covert supremacy whose companion, a white man, was equally exercised by the set-up. Companionship in naming the appearance, confessing the effects, and themselves laughing about the near ubiquity of white complicity in continuing supremacy, in this case, included someone white-skinned! White people who shift locations socially and politically, hooks asserts, are also able to see the way whiteness functions to terrorize without themselves feeling locked into denial or guilt (hooks, 1992, 177). Such an experience, for her, does herald hope: the possibility of a dissociation of whiteness and the terror it so covertly and concretely enacts—an exorcism of supremacy from white identity!—that names the content of any effort at modern decolonization.

Henry Louis Gates, Jr. and Signifyin(g) Tongues

hooks in effect extends the analysis of Long with everyday experience, naming whiteness as a myth of benevolent and benign sameness that hides its actual history as terror. She too invokes the language of ritual, constructing out of her own activity the possibility of understanding her daily struggle as mythic combat, keeping it linked with and energized by memory of other struggles, other combats, both won and lost. Her weapon is an "oppositional gaze," ever probing for the misuse of power, whether by white or black, naming the wound, gathering community, exploring alternative possibilities, before locking a word into the barrel of the tongue to fire back. And in her effort we meet a generosity similar to Du Bois, mirroring recognition that white eyes also harbor fear, that white bodies too may move in time and space to recover the *pathos* of power, to return domination to its crime, to confess rancor, to reinvest imagination with a confrontation that frees. The laughter enjoyed by hooks with her newfound conference friends (noted above) was sparked by hooks's asking how the woman, after battling every day with all the little rituals of insult and tiny gestures of terrorism, can "deal with coming home to a white person?" (hooks, 1992, 177). "Oh you mean when I'm suffering from White Person Fatigue Syndrome?" she cracked. "He gets that more than I do." This is a tactic that literary critic Henry Louis Gates, Jr. would label "Signifyin(g)"—a means of deflecting and defusing "put down" by way of "put on."

Here we shift emphasis from the quality of a look to the pirouette of a pun. Gates delineates the finesse in his 1988 book *The Signifying*

Monkey, which names "Signifyin(g)" (capitalized, with the final "g" in brackets to designate it as a vernacular form of speech) as a communal pedagogical practice that reconstitutes oppressed American black people as *homo rhetoricus Africanus*(Gates, 1988, 75). For Gates, the overarching sign of this black rhetorical tactic is a story figure known as the Signifying Monkey, immortalized in the twentieth century in various jazz riffs. Its origins go back to slave tales of a voluble trickster whose likely provenance is West Africa (and the traditions there surrounding the messenger-god and spirit of the crossroads and of choice known as Eshu or Eshu-Eleggua). In substance, the various versions of the story detail the triumph of this dissembling Monkey over a strong lion who is unable to decipher the difference between figurative and literal speech. The Monkey proves his mettle among the jungle's larger beasts by provoking the lion to initiate a (losing) fight with their mutual friend the elephant by repeating to him insults supposedly generated by the elephant. In fact, the Monkey's repetition is really only figurative. It is a matter of "signifyin(g) on" the lion's outsized, king-of-the-jungle pride, setting in motion the lion's own miscalculations of power and intent that lead to defeat. As a trickster figure, the Monkey incessantly relies on his ability to mediate between forces stronger than himself, whose opposition and then subsequent reconciliation he engineers through his capacity for tropic manipulation and reconfiguration. Gates points out that such "(anti)mediations" have the effect of both inducing and domesticating conflict (Gates, 1988, 56). The pattern established is not simply oppositional, but complex. Not only is the grandiosity of the lion deconstructed, but also at the end of the episode, the relationship between the elephant, the lion, and the Monkey is reconstructed. The Monkey ultimately seeks not just triumph, but community.

But if we focus only on the story referents, we miss much of the point here. Gates is concerned not so much for a disappearing story as a living practice. While relatively few contemporary black folks are "accomplished narrators" of the Signifying Monkey tales, "a remarkably large number of Afro-Americans are familiar with, and practice, modes of Signifyin(g)" (Gates, 1988, 54). At first glance, the Monkey tales offer themselves somewhat transparently as an allegorical hermeneusis of the racial conflict between white and black in America. But their triadic structure (lion, monkey, and elephant) resists that easy binary referentiality and redirects attention to the constant emphasis within the tales themselves "on the sheer materiality, and the willful play, of the signifier itself" (Gates, 1988, 59). It is the Monkey himself, "in himself," who is the master-focus of the telling. And "in himself" he is "not only a master of technique . . . he *is* technique, or style, or the literariness of literary language: he is the great Signifier" (Gates, 1988, 54). And it is in this sense, says Gates, that we discover the "true" import of the stories: "one does not signify something; rather, one signifies in some *way*" (Gates, 1988, 54).

"Signifyin(g)," for Gates, signifies the operation of black creativity in the "break" between black and white worlds. As designating a black speaking practice, it points to a skill directed at multiplying meaning in the gap between the signifier and the signified. At the level of the signifier itself, words taken over from English generally retain their standard meaning. At the level of the signified, however, black speakers have been able to load in a range of additional meanings very specific to black culture. In so doing, they have opened up a parallel universe alongside white discourse (Gates, 1988, 45).

To "signify" on someone, in black meanings of the term, is to "trope" them, to take up their own communication and repeat it with a signal difference that simultaneously highlights its origins as creativity and outcreates that creativity by forcing it into new levels of meaning (Gates, 1988, xxii). Formal linguistic analysis of the process reveals a nimble shifting between syntagmatic and paradigmatic "registers" of meaning. English-language use of *signification* generally focuses on the way meaning is a function of highlighted resemblances. It implies a "chain of signifiers that configure[s] horizontally, on the syntagmatic axis," where each word points toward another in a sequence of synonyms related to each other in their similarity of (standard) meanings (e.g., "tea" means something like "stimulant" means something like "caffeine" means something like "coffee," means "hot beverage," etc.—similar to the discussion on mean-making in chapter 4) (Gates, 1988, 49). *Signifyin(g),* on the other hand, subtly emphasizes differences. It conjures its playful puns in the mode of homonymic substitution, where, rather than finding dissimilar words related by obviously similar meanings, we find dissimilar meanings gathered in proximity by their similar sounds. Such a practice brings into play "the chaos . . . of Saussurean associative relations" that underwrite meaning . . . along the vertically suspended, paradigmatic axis (Gates, 1988, 49). It probes all the excluded possibilities underlying the syntagmatic register ("tea" might set off an associative play of thought running to "T" to "T-Bone steaks" to "Ice-Tee" to "rhapsody" and "rap city" and "graffiti"). Incisive humor, then, not syntactical propriety, rules the roost of black figuration. The *savant* of street-signifyin(g) will name persons or situations in the "telling manner" of a sharp eye and sharper wit, luxuriating in the "free play of [the] associative rhetorical and semantic relations" constraining the unconscious of the dominant white order (Gates, 1988, 49). Linear coherence is here superseded by rhetorical concatenance.

For Gates, then, the heart of the practice of Signifyin(g) is its ironic refigurement of the boundary between black and white linguistic universes as the structure of a *homonymic pun* (Gates, 1988, 47, 75). The white/black interrelationship that we have previously characterized in terms of the Du Boisian double is here translated into linguistic analysis as a "vertiginous symbiosis," painfully opposed in its interdependence, playfully ironic in its simulated identifications. Signifyin(g) offers its career as a "redesign" of the sign, an apparent sameness that hides a profound difference.

This masking function (hiding a difference) is, in part, definitive of the practice. Historically, the black community has had to learn adroit manipulation of language as a basic survival skill. Gates underscores its power even up to the present. Teaching black children "smooth navigation between these two [discursive] realms has been the challenge of black parenthood" (Gates, 1988, 76). But ironically, then, the "fine art" of linguistic circumnavigation between two worlds at the same time functions, in effect, as an alternative language, peculiar to and partially constitutive of the second (black) community alone. Away from white gaze, "black people created their own unique vernacular structures and relished in the double play that these forms bore to white forms" (Gates, 1988, xxiv).

In scholarly circles, focus on the survivalist aspect of black tropological pedagogy has meant almost exclusive attention devoted to the gaming practices of "capping" or "sounding" or "playing the dozens" (Gates, 1988, 79–80). In these latter forms of verbal dueling, predominantly male adolescents train younger adolescents in forms of "put down" that generally attack the family line ("your mama") and test quickness of repartee and ability to absorb abuse without giving way to anger. The emphasis, here, has fallen on rituals of insult that express aggression as a form of catharsis.

Other scholars have pointed out, however, that Signifyin(g) is, in fact, both a much broader concept and a much wider practice. It encompasses at least twenty-eight different rhetorical "sub-tropes" (talking shit, woofing, spouting, muckty muck, boogerbang, beating your gums, talking smart, putting on, telling lies, shag-lag, shucking, jiving, rapping, etc) and is common to adults, not just adolescents

(Gates, 1988, 77). Its province of operation is not just that of masked communication dealing with, and displacing, verbal assaults coming from the intruding white world (indeed, much of the innuendo that Signfyin(g) depends upon for its effect would go undetected in white ears). Rather, its generative institutions are indigenous "black" theaters of practice (the home, barber shop, hairdresser, etc.). It is, at root, an intra-*black* practice of language-users so delighted by metaphorical play as to warrant description as "the community of the 'Third Ear' "[1] (Gates, 1988, 76, 70).

In specifying the nature of this community constitution, Gates details the ubiquity of the practice. Signifyin(g) can be used to put others down, build them up, or simply express one's own feeling. Its means encompass not only spoken figures of all kinds, but seemingly inarticulate moans or "unmeaning tunes" and a whole range of bodily gestures, from "signifying eyes" to "talking hands" (Gates, 1988, 67, 69). Its significant features are perhaps best referenced as "implication" and "indirection" (Gates, 1988, 75, 77). For its effect, it can require that its hearers detect its appearance (amused awareness that one is being "signified upon"), or in other cases, rely precisely on lack of detection to accomplish its "devious" intention (eg., the lion in the Monkey tales) (Gates, 1988, 85). It emphasizes behavior as performance, speech as gesture, personality as "impersonation-with-a-difference," learning as improvisation, relationship as agonistic play, encounter as "ad-lib quickness," and success as "the coaxing of chance" (Gates, 1988, 76). Within its province of operations, rehearsal becomes a way of life. It is finally a "formal device of style" that defines a rhetorical blackness of conversational being (Gates, 1988, 85, 84).

Kristin Hunter Lattany and Off-Timing Mockery

Deployed as a deflection of racial domination in all the daily spaces of black–white encounter at work, in school, at the club, Signifyin(g) stands as a sign of delirium, a possession of language itself by energy under duress. The finesse in view is quite capable of reconstructing the time and space of the encounter in a mere flick of hand or cock of head over fevered, or slow-as-molasses, word. In the confrontation with such, white speech—like the lion in the Monkey tales—often stumbles unsure, knowing itself dazzled but unclear on the meaning of the bedazzlement. It is "signified upon," caught in a dizzying proliferation of ironic or sardonic references that require "insider information" for their exact decoding. Such a display sets in motion an alternative "tactic" of power—in the manner in which Michel de Certeau differentiates ad hoc "tactics" from more systemic "strategies"—based not on position, but on seizure of a moment, takeover of a space, for only as long as the interaction lasts. The resulting dis-ease in the white belly, however, is a symptom of war, and a sign of a battle just lost. But such experience is necessary for supremacy to begin to recognize itself consciously and is fundamentally a demand for the reciprocities of shared rhythm rather than the hierarchy of control, as novelist Kristin Hunter Lattany argues.

What Gates has theorized as Signifyin(g) and underscored as a survival skill, Lattany specifies as "off-timing" and more prophetically assigns an apocalyptic function. Like Long and hooks, Hunter peeks under the Veil ghosting white America in glitter and allure, and finds not brightness but a skeleton (Lattany, 164). She is concerned to name the thrall and reclaim her own fascination by tracking black resistance tactics over the previous century as they changed with each new generation. For instance, the period embracing those in her community who benefited from civil rights and integrationist struggles produced, she says, "the first African Americans . . . to have a purely white consciousness" (Lattany, 166). Born in the 1950s and 1960s and socialized into the Yuppie era cares of the Bush and Reagan years (the 1970s and 1980s), these "favored ones" inherited rather than fought for placement in white schools and suburbs, corporations and professions, and aspired, with hair permed and personalities bleached, to offset their color with possessions (Lattany, 167). It is an aspiration Lattany is concerned to repudiate. Historically situated between the savvy and fiery Soul People who preceded them and the brave if foolish Rap Generation following, this Rock and Roll Generation, she claims, is afflicted with "soul loss," destined to dissatisfaction or spiritual death, in the attempt to disown black difference as mere accident.

But she herself almost succumbed—until shocked down to her roots, one day in 1986, when gathered with her family in Georgia around TV coverage of the space shuttle *Challenger* launch as it blew up. Unmoved by the tragedy and trading comments on the hubris of the name and the wantonness of the aim to colonize yet another space before learning how to live in this one, the family had suddenly "coiled" as one body in alert interest when the photo of the lone black member of the crew was shown. The posture itself rebuked, springing the author back to her own upbringing in the 1940s and the training in "off-timing" and "being a race woman" she had so enjoyed and nearly forgotten. The "clear distinction between *theirs* and *ours*" immediately re-presented itself and demanded choice, Lattany recounts, and that choice remains "a darkly hidden and [recurrently] necessary part of being black in America" (Lattany, 174). In bitter knowledge lies strange liberation.

Fundamental to this in-group recognition is apocalyptic evaluation of American world-domination as a death-cult. Lattany does not mince words, nor does she elevate Africanity as some kind of innate immunity in the charge (she criticizes the continent with cooperation in slavery and repression of women, among other things). Rather, she notarizes a historic "African" difference of more "naturalized" rhythm—black bodies moving in time with, rather than against, nature, at ease in themselves, eating greens, loving life, loving loving, affirming kin, revering not controlling the environment (Lattany, 172). To the degree white America had itself "gone African" in the twentieth century in dancing to Little Richard or listening to Martin, ragging with Joplin in the Jazz Age or trying to talk "street" in the beats, "there was hope for America's redemption" (Lattany, 164). But the more recent emphasis on war and weapons-love, the fascination with fluorocarbons and nuclear wastes, liquor and drugs and the taste of suicide, the flattery of a society rushing like lemmings off a cliff in its battery against the rest of the world—all these she sees as a "great white death urge," sweeping everyone who succumbs, white, black, yellow, and brown alike, into its mass grave (Lattany, 164–165).

Her remedy? The modality of off-timing—double-consciousness in creative action (whose split she would otherwise abhor if it were merely passively suffered). Off-timing, for Lattany, is

a metaphor for subversion, for code, for ironic attitudes toward mainstream beliefs and behavior, for choosing a vantage point of distance from the majority, for coolness, for sly commentary on the master race, for riffing and improvising off the man's tune and making fun of it.

The cakewalk was off-timing, mocking the airs of Massa and Missy, and making the antics of Philadelphia's New Year's mummers doubly, deliciously funny to blacks in the know. Louis Armstrong's mockery of minstrelsy was definitely off-timing, shored up by his strong talent, as it had to be; one needs a strong sense of self to play off the master and his stereotypes. Those Hampton students who coolly ignored George Bush's commencement speech in 1991 were off-timing. Haitians who adapt Catholic saints to voodoo rites are off-timing. No matter what she says, when Marian Anderson appeared on the Lincoln Memorial stage as a substitute for the DAR's—and opened her concert with "My Country 'Tis of Thee"—she was making dramatic and effective use of off-timing. The sister in Washington, D.C., who said, when asked her reaction to the Queen of England's visit, "Well, she's fascinating. But so am I," was definitely off-timing. Bill ("Bojangles") Robinson's tap routine to "Me and My Shadow" was a superbly elegant bit of off-timing, a mockery of the mocker that made a suave, ironic comment on racism.

Off-timing, I learned in my youth, was the subversive attitude we had to maintain if we were to survive in the man's society. It was Uncle Julius's sly devaluation of European values in *The Conjure Woman*, and Larnie Bell's knowing jazz renderings of Bach fugues in *God Bless the Child*, and was more a skewed, but single, ironic consciousness than a double one. (Lattany, 165–166)

To the degree black double-consciousness represents a loss of contact with the inner self, a parroting of self to parade likeness to whiteness, Lattany rails against its disingenuousness. But when it amounts to dexterity—"using a white voice to talk their talk at work and slipping fluidly, instantaneously, into the race vernacular one minute after quitting time"—it is biculturality serving survival (Lattany, 172). Here, Du Bois's doubled awareness slides into active tense, enabling, when one is forced to dance with Death on a daily basis, a syncopated self-maintenance, dipping in and out of time with the mainstream beat, keeping "cool" in spite of the heat and madness. And at the receiving end of such, on the side of things designated by Lattany's off-timed phrase "white folks!" punctuated with shrug and eye-roll, the mockery is potentially delivery in the form of a raised eyebrow. The death mask either splits in a self-revealing—and self-recognizing—grin or hardens into cold violence.

Paul Gilroy and the Slave Sublime

Lattany makes us aware of a mobilization of double-consciousness as counterterrorism (as well as of its potential for self-dissimulation). As with those cited before her, she recognizes her wrestling is with a form of madness, a drum beat of death, a march off the map into annihilation. The American fixation on bars and bombs, on hallucination and conflagration, she names with urgency and near despair in her voice. The choice to integrate with a burning house, as Malcolm raged, is no choice at all. But she also nuances the tactic to tryst with the tyrant when necessary. White supremacy gives birth to sycophancy, and to suicidal bravados of resistance, both. Between such, between herself and the other world, however, there is also the echo of another possible step, born of polyrhythmic practice, the interweaving of imitation with novelty, moving off key in doubled time, signaling savory mockery to those who know the code. Making of a tiny gesture a world of difference and laughing in the suddenly gathered "big body" of friends and ancestry—this is a corrosion that white control has never been able *to* control. It is a structure of celebration cultural critic Paul Gilroy names "sublime."

Where Lattany might seem to border on a kind of essentialism of black difference, Gilroy waxes prolific in calling for new theoretical tools capable of doing justice to the hybridized expressive forms of diasporic African cultures. For Gilroy, too, the central trope is the Terrible: the slave-trauma found in an intensity of black memory that is actually conserved and cathected rather than eliminated. But for him, the out-working of such in the ritualized dramaturgy of black culture is not to be analyzed as the expression of some essential core of the culture but as ever innovated in response to the pressures of modern capitalist assaults on life. Gilroy is concerned to theorize what is common to black Atlantic cultures without essentializing a "thinking racial self" or a "stable racial community" (Gilroy, 1993, 110). He pushes theory past its focus on genteel notions of beauty and taste to the vernaculars of mimesis and performance, to the codes of a "lived racial sense" whose forms of stylistic appeal and sedimentations of secret meanings can be both taught and learned (Gilroy, 1993, 109). The move is complex and concentrated, zoning in on micrological gestures of the body rather than narrative meanings of texts, privileging music as the preferred modality of memory and resistance, and intensity of feeling as the sign of a shared structure of survival. Gilroy ranges wide in his solicitations, moving from the terror of slave situations to the titillation of blues intonations, from social movements of the urban dispossessed in Britain to the svelte funk of Toni Morrison's prose, but continuously circling round to the question of the "willfully damaged signs" of the "slave sublime" (Gilroy, 1993, 37). By these are meant the "screams, wails, grunts, scatting, and wordless singing that appears in all these black cultures"—the "distinctive kinesthetics" of a diasporic poetics of the body—indicating that here there are "meanings and feelings so potent, so dread, that they cannot be spoken without diminution and trivialization" (Gilroy, 1987, 75, 212). These registers of mutated communication, of deformed signification, cannot be grasped by European canons of "the rational." They are utopic manners of expression, bodies laboring the duress of enslavement and colonization, economic exploitation and social enghettoization, into recognizable forms of community congress and protest, retrieving the black body from its submergence in the commodity form and mobilizing it in a "dialogics of consumption" that bridges art and life (Gilroy, 1987, 164; 1993, 37).

For Gilroy as well, Du Bois is pedagogue and partner. *Souls* emerges, for him, as a "self-consciously polyphonic form," combining "recognizably sociological writing with personal and public history, fiction, autobiography, ethnography, and poetry" (Gilroy, 1993, 115). Here too (in the genre itself) is expressed recognition that no one form alone was adequate to the intensity of feeling engendered in the history of slavery and its aftermath. And here also is testament to the improvisational character of the culture explored: Du Bois is candid about "the way he had to learn the codes, rhythms, and styles of racialized living for himself" (Gilroy, 1993, 116). Du Bois looms as the intellectual who clarified for Gilroy the "constitutive role of terror in configuring modern black political cultures" and who linked the codification of that history of terror with the expressive particularity of black music (Gilroy, 1993, 118, 120). In *Souls,* says Gilroy, music for the first time is employed as "a cipher for the ineffable, sublime, pre-discursive and anti-discursive elements in black expressive culture" that give rise to a "diasporic multiplicity," a "tradition in ceaseless motion," a fractal morphing of cultural forms in which "the relationship between similarity and difference becomes so complex that it may continually deceive the senses" (Gilroy, 1993, 120–122).

And like Long, Gilroy traces this polymorphic synergy to the spiritual practices of black religion in which the "buried social memory of the terror had been preserved" and was regularly revisited by ritual means (Gilroy, 1993, 129). But Gilroy also traces a genealogy of this memory through the quite varied and continually changing black Atlantic musical cultures that are its prodigious issue and offspring. Du Bois's double-consciousness here marks a rupture that is taken up in musical venues as the tradition of the "break," a kind of "utopian eruption of space into the linear temporal order" of "the regimes that sanctioned bondage" (Gilroy, 1993, 198, 212). The biblical Jubilee tradition (of sudden release from long-term servitude) is blasted out of its merely textual orbit in these musical cultures and re-anchored in the black body in "blues" motion (Gilroy, 201, 212). Inside the hell of slavery, or the desperation of Jim Crow tenant farming, or the despair of industrial and postindustrial menial labor, this body becomes simultaneously communal and virtuoso, a "soul" of improvised flesh, moving to a "lower frequency" politics, animated by the energies of a remembered terror that becomes the throbbing artery of self-recognition and communal affiliation for those who move to its beat (Gilroy, 1993, 37, 79, 102, 202). Here, whether we focus on solemn spirituals or ribald blues, smooth soul or rancorous jazz, delirious dub or a rabidly rhyming rap, sacred ritual and profane performance can be observed in nearly identical effects: the institution of a constituency of active listeners, gathered around the priority of the present, in a form of agency dissolving differences "into the sublime and the ineffable" (Gilroy, 1993, 203). One of the differences sometimes and in some places (like reggae clubs and rap concerts) "dissolved" is that of white "arrhythmia."

Tricia Rose and Rap Rupture

Gilroy sounds much like Long and hooks in his fascination with the *Tremendum* of black experience, recognizes fully the facility with language championed by Gates and Lattany, but finally rests his own enthrallment with the body in communitarian syncopation, rocking to juke joint joy or a midnight growl of blues-pain at the crossroads of destiny and damnation. It is the body in which all of the remaking of time and space, timing and placement, theorized by the others finds its womb and home. This is a spirituality of funk, of sweat under the duress of spirit, labored with memory, breathing like the end of time, calling up "that slavery shit" from the bone into a muscle in motion, working the anguish into ecstasy or into the discipline of a mind like Ralph Ellison or Zora Neal Hurston. This is the whole of modern history in a single long scream of James Brown, outside the word, eloquent in erudite gesture and carnal antiphony. The cosmos-become-God-quake that Long labels *daemonic* in colonial relations, the carnival of castration and lynching that hooks engages in a hard look, the "jungle wit" of the tongue of the Monkey, the off-beat sweetness of Lattany's mockery—all of these in Gilroy find their closest synonym in the hard knot of intensity that he finds at the core of expressive black music. But this remaking of white repression into insurgent rhythm is a prodigal vocation. It finds one of its latest incarnations in the hip-hop innovation cultural historian Tricia Rose unpacks as "ruptured flow."

Rose's 1994 reprise *Black Noise: Rap Music and Black Culture in Contemporary America* opens an articulate window on the intersection of "social alienation, prophetic imagination, and yearning" that is the postindustrial ghetto now rendered *geist* of the hour in global musics (Rose, 21). Early "dissed" and dismissed as a mere trend headed quickly out of sight around the bend of history, hip-hop culture has emerged over more than a quarter century now as the new cipher of global youth culture up against the wall of impossible demand. Late capitalist reconstruction of adolescence in the image of the market, while simultaneously serving conservative interests seeking moral conquest over every form of young desire, has resulted in a profound dilemma. In the ghetto such a contradiction is imbedded in landscape and life alike, and young artists of the late 1970s— working out of the efflorescence of Caribbean and Bronx energies, new technological and old ideological possibilities, and their own impossible placement in a social order determining their destiny as "prison" or "grave"—crafted a new sonic probe of the desperate paradox. Rose rides the new rhythm into academic deva-hood with her work and has become herself a kind of prophet of the polymorphic tribe of digital trance and toasting that now enfolds a globe. The "word up" of her throw-down with the culture is a sharp and sultry sounding of the sound, a digging out of the percussive ground opened by rap rhyme, graffiti sign, and breakdance mime that uncovers an entire planet in its sonic depth. Rose herself is a child of the changeover and "checks" the feminist left's tendency to reject the movement wholesale as just more male misogyny. The result is a book and a career that complexly figures hip-hop's advent and surfeit, struggles and muddles, with a critical eye on the global scene that loves to consume the hard edge of black creative labor without engaging the politics that keeps its artists and communities face-to-face with denigration and early death.

At the core of Rose's reprisal is an analysis of the basic signal animating the primary forms of the culture (graffiti moniker, breakdance move, and rap music). She identifies hip-hop genius as a particular coalescence of what she calls *flow, layering and rupture* (Rose, 38). Whether in the shouting, oscillating syllables of the "tagged" building, claiming local rights to social space even in the face of disenfranchisement, or the popped and locked joints of the body made into a living mime of the machine, or the staccato rhyme over scratched lyric and sampled drum line, the outline of the shared values echoes through the disparate forms (Rose, 38–39). Rose argues that the depth-structure of the common valuing is a matter of style-as-war at the "crossroads of lack and desire in urban Afrodiasporic communities" (Rose, 35, 61).

Theoretically, the values could be said to

> . . . create and sustain rhythmic motion, continuity, and circularity via
> flow; accumulate, reinforce, and embellish this continuity through lay-
> ering; and mange threats to these narratives by building in ruptures that
> highlight the continuity as it momentarily challenges it. These effects at
> the level of style and aesthetics suggest affirmative ways in which pro-
> found social dislocation and rupture can be managed and perhaps con-
> tested in the cultural arena. Let us imagine these hip-hop principles as a
> blueprint for social resistance and affirmation: create sustaining narra-
> tives, accumulate them, layer, embellish, and transform them. However,
> be also prepared for rupture, find pleasure in it, in fact, *plan on* social
> rupture. When these ruptures occur, use them in creative ways that will
> prepare you for a future in which survival will demand a sudden shift in
> ground tactics. (Rose, 39).

While recognizing that these tactics are shared across a wide range of
Afrodiasporic cultures, Rose is also savvy in discussing their relationship to the commodity market and the exigencies of consumption
(Rose, 39–40). They do not so much directly contest as complexly
"confess" a differential relationship to the commercial establishment
that both propagates and reshapes their effects. Cut by continuing patterns of male dominance, exercised by continuing anger over racial
marginalization and class subordination, crossbred with technological
change, and inflected by concerns for comfort easily manipulated but
not so easily controlled by the market, hip-hop creativity remains a
strikingly subversive reflexivity of the postmodern moment (Rose, 61).
It encodes pleasure and pain, capitulation and contestation, in a style
of competition on the hard streets of the city that remain a primary
locus of "other" modes of American identification. That the culture's
primary American market is "crossover" has everything to do with
young white desire for a "quality of being" that is bigger than life
inside a Budweiser commercial and a suburban "hovel." As social artifacts, rap riposte and DJ-ed assault on soft living are not mere posture
(though they may have become such for many of the artists now conveying them), but conduits of a certain charge of vitality condensed
historically from real struggle against real constraint. White youth recognize the signal even if they have no clue about the condition that
produced it. More broadly, hip-hop percussion marks a continuing
agony at the heart of American imperial expansion, and hidden white
aggression, at one very deep level, remains a primary target of the
sonic probe. White confusion and fascination over the product do not

thereby negate the politics. Or the potential for troubling revelation, should ghetto rhythm be embraced as more than just entertainment.

Profiing Blackness

But it remains an open question whether theology as a modality is capable of adequately expressing the kinds of profound combat and nuance of negotiation just outlined. Theology's power as discourse has too often subverted its own emblem of power crucified. The insistence of Charles Long and bell hooks that the profound historical and cosmological rupture signified by "blackness" be faced as "tremendous" and terrorizing rather than merely tantalizing must indeed be faced. It must not first be tamed in a formalized system, or its very meaning is already lost.

The Space of the Tremendum

Long's insight presents us with the possibility that the empty space between the two signifiers of race harbors a hidden god. It asserts that black and white mark the site of a historical struggle that has imposed revelation on one side of the divide even as it has blinded on the other. The violence of the contact—the simultaneous rupture and (supposed) "repair" of the cultures encoded in the "order" the two categories impose—has had a religious effect and engendered a mythic response carried in a style and *habitus* of the body. Blackness, on the black side of the divide of race, is religious *sui generis*. It comes into being as an imposition of terror; it claims its being for itself as a community of vital resistance.

I begin with Long in this chapter on black performance because he begins before speech. The roots of blackness are not first articulateable, but only expressible. They germinate in the silence of a metaphysical shudder. They begin to live in history as a groan, like the ancient groan that launched the biblical story of oppression and its overcoming in Israel's flesh (Ex. 2: 23–25). History itself, undergone in a black body, is "initiation," a rite inscribing its heinous rupture, between the worlds of innocence and maturity, in blood. Race is not biological, but blackness *is* a thing that springs from the blood, the flesh, of a bleeding god. It is religiousness before religion; the sigh of the oppressed creature transfigured in the silent thunder of the offended Ultimate. Blackness here is the *mysterium tremendum* possessing a people against the cruel fascination to which they have been subject, possessing them as the remoteness that removes them from an aberrant white gaze. It is the splitting of night, the doubling of sight; Du Bois's "peculiar sensation" of ever "seeing oneself through the eyes of others" itself interrupted by *the* Other, author of identity and contingency, and of radical identity by means of radical contingency, doctor of the root, root of all doctoring and deliverance, the meaning of incarnation below the level of words, where the depths disappear into darkness, the adumbration-as-revelation of the black god—*in* the social body of blackness and *as* the aesthetics of a terror endured. Blackness, then, in one counter-meaning of the white stereotype, is an ongoing operation of divination, the improbable discovery of divinity in calamity, the making of life in the space of death.

The Timing of Signifyin(g)

But even within speech itself, ritual can encode a rupture that couples meaning with something else. Within the rhetorical privileging of the signifier found in Signifyin(g) and off-timing, an economy of speaking comes into being that does not seek to get anywhere in particular. Its payoff is not a surplus of information, the securing of something outside the signifier itself, but an experience of play and the creation of a commonness of being in space and time. Rather than a kind of accumulation, a halt of meaning to get at the "real," a swelling of knowledge, the very structure of black vernacular stages a display of rhetoricity and personality. Rather than the exchange-value of the "truth" it can refer to and "buy" from without, Signifyin(g) foregrounds the use-value of words, their magical possibilities of altering time and space. Language here, does not seek to slow objects down in space as much as speed the subject up in time. It creates, out of literally nothing but "empty" time and space, a theater for either a dazzling display of the human word or a subtle corporealizing of human meaning, but in either case, reveals a domination of signification by its community of creation. Here, at one level, is another form of Hegel's "revenge of the slave via mastery of the object": the object mastered is the space-time of the body itself, through a signification that materializes as style. Here also is a peculiar instance of Kant's ethical "treatment of others as ends, not means," the nexus of a proximity that does not objectify the person in the interests of trying to "get somewhere." Rather, Signifyin(g) and off-timing invite a profound subjectifying of personality within the least forceful means available: the invisible but leveling network of ambiguous rhetorical possibilities that can never be closed in a violent appropriation. While it would be foolish and unsustainable to claim for black-on-black interaction anything like an ethical superiority compared to other ways of interacting, the structure and discipline of Signfiyin(g) and off-timing practices themselves do incarnate an ethical predilection. They "enculturate" a force field of antiphony that gives "the other" a constitutive priority.

The Body of Sublimity

But in contemporary America the Word-made-flesh gains one of its deepest incarnational meanings not just in black religion and black speech, but also in a black body coded to live against the designs of death. The blues-body of black performance points to a profound meaning of God-on-earth. It points to a self-signified difference that must be grasped under the skin, in the unreachable skeletal-structures of cultural style and historical repetition.[7] It shows itself as the possibility of a subtle variation that emerges in those reiterative-codes of bodily style where the individual-in-oppression indeed repeats, but also reproduces-with-a-difference, the community-under-constraint. (Where the feet of James Brown both invoke and revise the memory of the feet of the Nicholas Brothers, for instance, or the "I mean business" gaze of Shaquille O'Neal's mother in the midst of a TV commercial can be easily re-imagined in light of the powerful eyes of an *ashe*-filled ritualist of West Africa.) The Afrodiasporic community did not only transvalue terror into collectively valorized forms of creativity. It has also served as the chorus for a wide range of individual improvisation and experimentation. It has done both, however, in relationship to the one "canvas of representation" (as culture critic Stuart Hall argues) over which the black community has had some control as a form of "cultural capital": its own forms of collective embodiment (Hall, 27).

Theologically, such a focus points to an understanding of incarnation that insists sacral difference is always at stake in the encounter of difference on the human plane. The negotiation of otherness within and between various human communities is always fraught with an uncanny-ness that can be rendered divine or demonic, healing or damning, source of liberation or excuse for destruction.[8] But in hope and history, that encounter hints at the sublime. It will not countenance separating spirit and body. In the case at hand, the conjured and conserved memory of slavery makes salvation practical in an imaginative rejoining of the arts and life, dramaturgy and death. It has import for embodiment *writ large*. Salvation is not only of the soul. It is in the body.

III

Presumption, Initiation, and Practice

From a theological point of view, reading black performance as (in part) a tactical response to white supremacy suggests placing in juxtaposition two histories of the body that can be made to question each other provocatively. Here, we would set Du Bois's articulations alongside the biblical text of crucifixion: *a hyper-tensed black body* nearly torn asunder by its doubling in the eye of race that continues to query American theology in particular as a "practice of privilege" and a *ruptured Jewish corporeality*, pinioned to the cross at the crossroads of empire and oppression, that continues to query Christian theology in general as a "practice of order." Two bodies, two bloods, two histories of anguish and its overcoming, as a task of articulation.

In framing the question thus, however, I am not only concerned with analyzing the black body as a resource for struggle and formation in the African American community, but also with halting before its complex significations as a sign of the entire nation. Specifically in this project, I am proposing the collective (and contested) performance of blackness in all of its manifold meanings, as a kind of critically productive *aporia* for the confession and reformulation of Euro-American self-understandings and practices. White America is put radically *in* question—corporeally and materially—in the very same instant the black body is solicited as revelatory (as it is in James Cone's polemics). In the historical logic of American racism, it cannot be otherwise.

But then we also need to go further and acknowledge that the spiritual polemics of a black nationalist group like the Nation of Islam strike close to the mythic core of America when they castigate the white system as "blue-eyed devilry." Any attempt to articulate the theological meaning of the white American body must begin with its likely condition as already "possessed."[1] The very material position and practical conditions of my identity as white are already in thrall to the real history and mythic fantasy of white supremacy, before I ever become either conscious in my intentions or deliberate in my actions. As a form of taken-for-granted identity, whiteness stands forth in the time and space of this country as a structure of violence and a significance of injustice. The black body as a "possibility of theophany" would thus have as its unrelieved mythic correlate the white body as a "question of exorcism." Theologically, the relationship between them must be understood, at one level, as a form of spiritual combat, a struggle for possession and counter-possession in the various times and spaces of both the body politic and the personal body. *Deliverance of either requires releasing each to a merely human and fully humanized, negotiation of difference.* Ultimately, exorcism of the powers of racism and healing of racial division demands a radical rethinking of the "*daemon* of difference" in relationship to the body incarnate. Such will be the preoccupation of chapters 6–8 under the rubrics of *white postures* (of being), *white initiation* into the racial situation (of America), and *white post-supremacist practice*.

6

White Posture

Then always, somehow, some way, silently but clearly, I am given to understand that whiteness is the ownership of the earth forever and ever, Amen.

—W. E. B. Du Bois (Writings by W. E. B. Du Bois in Periodicals
Edited by Others, Vol. 2, 25–26).

It began for most of us as vision, voiced-over in a commodity form, selling news. We were positioned as consumers. The tape recorded the aftermath of an interdiction: a black body stopped for speeding on a public highway by three sets of policing forces charged with overlapping, but differing functions of protection. The highway was the taken-for-granted national "frame"—a piece of public sphere, constructed as part of the federally subsidized post–World War II differentiation of metropolitan space into zones of wealth and want, transparency and color, caviar and criminality; serving initially as link between commercial "downtown" and suburban "bedroom," but increasingly in recent years shuttling resources, power, and appropriately inscribed bodies between the various epicenters of the new suburban hubs of American prosperity and identity. This particular body had been circulating not in South Central, but in the (suburban) neighborhood of Lake View Terrace, in a time (post–December 1987) and place (near the posh neighborhood of Westwood) when memory was still fresh of a young white woman who had been caught in the cross fire of a drive-by-shooting (Dumm, 184). The body had gotten out of the car, dared dance, and swiftly met the end of its mobility at the end of a baton. Judged excessive in its "timing" (a "foreign body" moving too fast through a "quiet space") it was made subject *to* the space it had "violated": a body, reduced to pavement, supine, flat, absorbent. It was not a singular strategy.

Later, in the spaciousness of a Simi Valley courtroom, under the time protocols of the docket, the ritual was taken up afresh and repeated, reinforcing the covert inscription of another space-time configuration as "white." In this case, it was a foreign use of technology that had to be reduced. The eyeball of video, producing documentation of an "excess closer to home," itself had to be exorcised. Its evidence of movement "out of control" (the 56 blows in 81 seconds, plus numerous stun-gun blasts, and boot kicks) was likewise squeezed back inside the regnant norms of enclosure, under the discipline of defense lawyer "scissors and pins" inside the courtroom. The body of tape—like the body of skin—must be staged to reveal the right "spirit."

The staging was not innocent. The courtroom strategy of the LAPD defense attorneys amounted to a loaded technique of *disaggregation*: a shrewd shredding of context until the video was finally pinioned, freeze-frame by freeze-frame, across a "clean white illustration board" (!), in demonstrably vain search for the single picture that would evince "excessive force" (Crenshaw and Peller, 58). Isolated from the whole, each single "still" could be "credibly" re-narrated in accord with published police protocol such that every blow, completed by its technical term of art, could "justify itself" as reasonable restraint. An operative norm of objectivity, informed by a taken-for-granted impulse to isolate and individualize, resulted in prejudicial sleight-of-hand: under the guise of "speaking for themselves" naked "facts" were *made to speak* a point of view.

But the question here cuts deeper than mere distortion. The search for objectivity itself already admits an accession to ideology (Crenshaw and Peller, 59). The popular presumption that the video represented clear, unambiguous evidence of a violation of civil rights also concedes too much. "Brutalities" and "violations of rights" are never self-evident, but invoke narratives. The sense, here, that the video represented something unmistakably damning, already itself subtly invests the rule of law with prescriptive narrative power. In that same moment, rights are accepted as a matter of formal calculus, proof as a matter of objectivity, objectivity itself as a surface effect of vision, and the "subject-on-trial" as the circumscribed individual.

Judgments of police brutality then cease to be a question of the relationship between masses of blacks (or Latinos) and the LAPD, and become a matter instead of satisfying the law. Racial considerations are radically thinned out in a formal structure of presumptive equality, symmetrically constructing its legal "objects" as color-neutral individuals. And in the very same act in which the tape's visual power is hailed as seemingly irrefutable in its objective meaning, the "merely verbal" testimony of thousands—unsupported by video tape—is necessarily marginalized as only "subjective" (Crenshaw and Peller, 66). The combat of language becomes irrelevant before the certainty of vision. The eye-that-judges has need of nothing more than what it sees. And what remains powerfully invisible in this "scene of seeing" is whiteness.

Putting Whiteness on Stage

What is whiteness? Most simply put, whiteness is a power of opposition. It emerges historically as perceived difference from, economic exploitation of, political dominance over, and presumed social superiority to, peoples "of color." In shorthand form it is the racial inverse of blackness, a negative conviction that "whatever else I might be, at least I am not that, not black." But it is a strange form of opposition. In listening deeply to Du Bois on the question of race, I would argue that doubleness can be detected wherever race is referenced. Not blackness alone, but every identification of racialized difference depends upon comparison and contradiction, repudiation of "the other" in service of affirming oneself. But whiteness at one level is the exception that proves the rule. It has functioned historically in mainstream America as a *de-racialized* position (at least in its middle-class formations) *vis-à-vis* the blackness it has marked out as the quintessential meaning of race. The "white" suburb, for instance, is not simply the "black" ghetto under a different sign of color. Rather, the oppositional parity the two terms seem to codify is an ideological subterfuge. Whiteness is, in fact, a very peculiar kind of opposite—a position, a privilege, a presumption, a pride, a propertied entitlement, a protected comportment, a way of walking, talking and "being" that operates not simply as an equal and inverse form of the thing it differs from, but rather precisely as its supreme judge. Whiteness here is not so much one term of a comparison as the eye that compares in the first place. And like any eye, the one thing it cannot see is itself. It is rather, for itself, a strange form of invisibility. The work of this chapter is that of outing such invisibility by questioning its (near) ubiquity in this country.

To ask about race in late-capitalist America is not simply to ask a question of the intentions or the heart. Much more consequentially it is to ask a question about power—power to control wealth, power to control discourse, power to control institutions, power, finally, to control public and private space and the codes of normativity and legality that "police" those spaces. The primary question the situation of race addresses to white people today is not one of interpersonal reconciliation, nor even only of white racism, nor finally even of white supremacy. Each of those descriptions leaves the problem as an adjective, on the "outside," as one of mere surface—a problem, one is tempted to say, of mere skin. Race is not a problem of skin. It is a problem of the body, of its place of dwelling, of its source of nurture, of its social scripting, its educational training, its resources of protection, its erotics of desire, its politics of control, its ecology of energy. It is a political problem with a psychological source ... or perhaps it is the other way around. But in any case, it is a matter of social constraint and cultural conditioning, as well as of personal caring, of horizons and habits, as well as of the heart. It is a problem whose remedy is indeed partly a matter of ritual, but not one whose theatrics can be choreographed in a single event, or even series of events. It is a problem whose root is as deep as the "self" we (who are white) think ourselves to be and as comprehensive as the country many of us move through relatively unimpeded. Ultimately it is a social dilemma with a spiritual genealogy.

Coming to grips with race as one of the forces shaping American identity today requires extensive work in relationship to both "time" and "space" in the United States—both the historical development of race and its institutional reinforcement. In its most difficult aspect, the problem today is whiteness itself. *Historically*, in this country as in others of our (old) New World, whiteness is theological if it is anything at all. But its coming into being was clandestine—the quiet effect of a noisy practice. Whiteness is the hidden offspring of white supremacy, which was itself the visible offspring of Christian supremacy. *Institutionally*, whiteness emerged as a "political buy-off" of other immigrant European identities that has since become the almost invisible presupposition of bureaucratic practice. Its very pervasiveness makes it hard to think about. It is so much a part of the institutional and cultural landscape as to be almost "cosmological," a quality of the very air Americans breathe. Both the genealogical work in history and the archaeological work on the contemporary social scene require extended treatment.

White Culture in the Making

Social scientist Ruth Frankenberg constructs a helpful framework by which to engage the historical genesis of modern whiteness and examine its complex legacy and profligacy. In *The Social Construction of Whiteness: White Women, Race Matters* Frankenberg offers three organizing devices that structure the reflections that follow here. She begins her work by delineating the "shape of whiteness" as a terrain of "linked dimensions" that encompasses "a location of structural advantage," "a standpoint . . . from which white people look at ourselves, at others, and at society," and "a set of cultural practices that are usually unmarked and unnamed" (Frankenberg, 1). She analyzes each of these three dimensions of structural location, subjective standpoint, and cultural practice in terms of (both) a set of material relations that concretely shape everyday experience and a set of discursive repertoires that serve to "reinforce, contradict, conceal, explain, or explain away the materiality or the history of a given situation" (Frankenberg, 2). And finally, she draws upon the work of Michael Omi and Howard Winant that divides the history of ideas about race in the United States into three distinct but overlapping stages (Omi and Winant, 15, 40, 69). Recasting their terminology into her own particular emphasis, Frankenberg typifies those stages as (1) a long-enduring paradigm of race-thinking based on perceptions of biological inferiority of people of color that she labels "essentialist racism"; (2) a period beginning roughly in the 1920s in which thinking shifted to "assimilationist" analysis that she comprehends as a projection of "color-blindness" attempting to "evade" perceptions of difference and realities of power; and (3) a post-1960s reassertion of difference in antiracist and cultural nationalist movements that she characterizes as "race cognizance" (Frankenberg, 113–115). While "whiteness" is clearly the quarry for Frankenberg's analysis, it emerges as an intelligible organizer of social perception and structure only gradually over a number of centuries. As an effect of complex historical processes, whiteness has gathered meaning to itself in ever-varying social and theoretical forms that can only imprecisely be separated out in my own thinking as "white Christianity," "white supremacy," "white racism," and "white normativity" (the latter three correlating roughly with the stages identified by Omi and Winant). These historical formations are the focus of the sections immediately following, before we turn to more phenomenological investigations of whiteness in terms of consciousness, class location, and "culture of desire."

Whiteness as Christianity

As an effect of colonial discourses of supremacist apology, white identity first began to take shape in history as an explicitly *theological* position (as we have already seen in chapter 3) (Dussel, 1995, 54–55). Whiteness was born of European encounter with people, places, and things that fit no clear category on the map of Christian cognition (Omi and Winant, 61–64). When Columbus and crew first began pillaging property in the misperceived and misnamed "Indies," the question of how to understand the indigenous peoples and aboriginal cultures he began to subjugate presented a conundrum. Natives did not immediately bow to the proclaimed dominion of Spanish king and queen, nor did they quickly embrace the God whose mission that dominion claimed to further (Wessels, 63). In the Spanish (and Portuguese) mind, such recalcitrance was "mapped" by means of discourse about salvation. The deep question was whether "Indians" possessed saveable souls, or only merely looked human but remained animal in reality (Dussel, 1995, 64–67).

The ambivalence of European experiences and approaches in the colonies percolated up to the theater of serious political debate back in Europe. As early as 1550, the king and queen of Spain adjudicated such a debate (known subsequently as the Valladolid dispute) between the Dominican "Indian rights" champion, Bartolome de las Casas and the Jesuit apologist for forced labor practices, Gines de Sepulveda (Dussel, 1995, 64–67). Although las Casas reportedly won the debate at court, he lost the battle for a different colonial practice in the colonies.

European "reading" of native practices (religious and otherwise) as radically "different" from their self-confessed Christian practices and beliefs served to license their will to dominate. The presumption that aboriginal intercourse with the world of spirit was at best humanely misguided, or at worst demonically so, pervaded colonial perception. The perception demanded Christian correction (if not subjugation or even annihilation) to return land and resources to profitable employment for God's "kingdom" (White, 160–164). In the process, European self-congratulation was ramified. The native resistance (inevitably) galvanized by the practices of takeover and domination only reinforced the sense of superiority presumed by the colonizers (Long, 1986, 123). Any possibility of religious parity or reciprocal spiritual learning between the cultures in contact was virtually unthinkable (with a few exceptions like las Casas) on the European side of the encounter. Even Jesuit attempts to create utopian communities of converted Indians (like the "Jesuit Reductions" in Paraguay or Brazil)—impressive as they were for humane treatment of aboriginal interests—were based upon a presumption of indigenous backwardness

(Dussel, 1995, 68–69). Christian triumphalism, freshly invigorated in the *reconquista* of Spain from Muslim "infidels," gained immeasureable momentum throughout the colonial period. Supremacy was a platform and a perception, not a question.

But the attachment of this superiority to skin color was a gradual process. The concourse between self and other, identity and difference, the subject of gaze and the object gazed upon, is always double, leveraging effects on both sides. European consciousness of light skin color as an immediate marker of "Christian" identity *vis-à-vis* native "paganism" not only invested bodily surfaces with spiritual sigificances. It also bled back into the religious symbol itself. Gayraud Wilmore, for instance, has traced the phenomenon of the "whitening" of Jesus in the European encounter with peoples of color (Wilmore, 228). In the same instant, this "pallorizing" of the Christ rebounded upon light skin. Modern Western Christian association of goodness and purity with whiteness, evil and sin with darkness, impurity with mixtures and off-colors, divinity with transparency, christology with European physiognomy, and soteriology with a progressive "enlightenment" gained its historical intransigence in this reciprocal process (Bastide, 273). The theological meanings invested in epidermal appearances served the function of theodicy, legitimizing the exploitation of both (indigenous) native and (imported) slave labor for European colonial and later imperial enterprises.

Whiteness as Supremacy

By the eighteenth century, the theological coagulation of "white" and "right," "light" and "might," began to gain ontological voice. As we have seen in chapter 3, in the aftermath of the breakup of Catholic Europe, anxieties about eternal destinies increasingly took on this-worldly flesh. Soteriological discourse about placement in the after-life found reincarnate "secular" focus in taxonomic discourse about placement in this world. Enlightenment philosophers and "natural" scientists supplied the map. In Hegel, for instance, Europe thought of itself as occupying the apex of a developing world-spirit, gazing back down the long reaches of time and out over the distant shores of space, tracing a German culmination of *geist* in the metaphysical arrival of self-consciousness at its absolute essence: consciousness of itself through its other. Here, Christian superiority climaxed in European supremacy. The "Great Chain of Being" organizing the ancient neo-platonic hierarchy of beings into a static universe presided over by the Trinity and ruled by Christian kings and popes in the Middle Ages was transposed into the modern key of a radically historical dialectic embodied in Northern European achievements.

Jorge Klor de Alva underscores the "achievement" embedded in this turn to ontology for the modern institution of slavery. Unlike many other forms of slavery throughout time and across the globe, the Atlantic slave trade had no clear legal rationale. Developed to supply African replacement labor for indigenous American populations decimated by war, disease, and ruthless work requirements in the colonies, this modern trade struggled to "justify" itself. In most places historically, slavery was backed by recognizable policy. People were enslaved for incurring debt, losing a war, committing a crime (Klor de Alva, 67). The recognition that enslavement was an accidental consequence attaching to a particular person rather than an essential condition attaching to a particular group remained extant (Klor de Alva, 66). Klor de Alva clarifies that the Atlantic slave trade, however, could claim no such possibility of legitimation—and in a long process of self-justification, had gradually evolved an understanding of African slaves as *theologically* created to be such and *ontologically* incapable of being anything else (Klor de Alva, 66). Where history could offer no ideological help, "nature" was made to stand in the gap. And where blackness was naturalized as the sign of a metaphysical constitution as "slave," the payoff back across the divide of color was a naturalization of whiteness as a metaphysical constitution as supreme.

In the United States and South Africa, in particular, this metaphysical "underwriting" linked up with a "Reformed sensibility" that tried to "divine" eternal dispositions by way of temporal indications (again, as we have seen in chapter 3). Initial rebuff of missionary efforts on the part of indigenous Africans was taken up, in Reformed anxiety, as the sign of destiny. Black skin reflected a "dark" heart toward God, indeed, a theological "curse" given epidermal figuration that was ultimately traceable back to Ham (Klor de Alva, 64). Christian superiority reinforced by metaphysical supremacy was re-reinforced by Calvinist indelibility. In this kind of "sign economy," white supremacy achieved its most virulent ideological articulation, as the inheritor of an absolute essence with absolute destiny.

In sum, as Cornel West and other race theorists have pointed out, white supremacy in the United States (and South Africa) thus emerges as unique in virtue of its complexly modern constitution (West, 1982, 47–65). It gathers force as an amalgam of (1) a political–economic institution of enslavement, (2) a phenotypical contrast of stark and seemingly unerasable differentiation, (3) a Cartesian self-consciousness crystalizing its identity in a unitary and individualized form of subjectivity claiming universal valence, (4) a scientific form of rationality seeking to prove its own transcendence by metaphysically categorizing the entire objective world (including dark-skinned human beings who were thought to be part of "nature") in a totalized taxonomy, (5) a Calvinist notion of predestination that sought eternal confirmations in surface significations (like success in business or skin-color in race), and (6) an Anglo cultural predilection that reacted to the color "black" with visceral horror and mental revulsion (Jordan, 7). It is simultaneously a science and a politics, a knowledge with terrible power.

Altogether, the coalescence of "scientific" racism in nineteenth-century America found its primal axis of contrast in a specifically Anglo-Afro confrontation that continued to exercise soteriological force precisely in its Enlightenment positivity. White ascendancy was its heaven; black enslavement its hell. Abolition and the Civil War notwithstanding, its career had just begun. Jim Crow, the lynch mob, and public segregation would be its southern offspring; enghettoization and pervasive discrimination its northern issue. In seducing immigrant populations toward its promise of race privilege and middle-class prosperity, "whiteness" would continually fracture working-class interests and thwart cross-cultural coalitions that could genuinely challenge overclass power. In the social history of this country "supremacy" has not merely served to code the dominant consciousness; it has also served to fragment oppositional movement.

Whiteness as Racism

But it is not primarily whiteness as (blatant) supremacy that emerges as a problem today. Such a formulation remains dangerous because it seems to promise the possibility that the supremacy can be carved off from the whiteness, leaving the latter intact as a viable form of identity. David Roediger, Noel Ignatiev, and Ruth Frankenberg help clarify the way whiteness itself came into being as a "commonsense" category providing cover for the operations of white racism. (In what follows, the term "racism" is understood and used, with one caveat, in Omi and Winant's sense of any racial project that "creates or reproduces structures of domination based on essentialist categories of race" [Omi and Winant, 71]. Their definition does not clarify whether such categories need to be consciously held and embraced or only mobilized "in effect" [even though consciously denied as in "color-blind" discourses of race]. The latter focus on effects rather than consciousness is the point of view taken in this essay).

Roediger's *The Wages of Whiteness* and Ignatiev's *How the Irish Became White* track the way whiteness was made to function politically in the nineteenth century to grant immigrant groups a "psychological wage" (borrowing from Du Bois) in lieu of real wages in the newly industrializing nation. As primarily a meaning of "not that," "not black," whiteness seemed to certify that "wage slavery" was at least not real slavery, not an obstacle to admission to citizenship in the republic as a free individual. Even in the North, where the encounter was one of "free" blacks with white (wage) "slaves," the categories are birthed locked in combat, pale-skinned Jacob leap-frogging into life at every turn by means of his grip on the heal of Esau the dark. (Jacob, if we remember the biblical story well, gains title to wealth and blessing from the patriarch only by using his brother's skin as a subterfuge for gaining favor. Modern-day Jacobs have merely inverted the gesture, hiding under their own skin to please "the father.") Promulgated in the political discourses of the Democratic and Republican parties alike, whiteness also showed up in the popular culture practices of minstrelsy and blackface celebrations. In these latter rituals of parody, white, primarily working-class, males blackened their faces and strutted their stereotypes . . . and then went out and beat up actual black people (Roediger, 106).

Roediger analyzes the contradictory behavior in terms of a psychoanalytic *topos*. In white male imagination blacks represented a pornography of the self left behind in Europe—a rural life of closeness with nature, familiarity with animals, freedom with the body—that was forcibly abandoned in the disciplines of industrialization and wage work (Roediger, 95). "Blackness" marked an intolerable loss in the choice to become "white." Immigrant groups in the process of "whitening" temporarily revolted against the restrictions that identity implied by "freeing" themselves in blackface, and then—lest anyone get confused that such was only play—reasserted their nonblackness physically.

Nostalgia may indeed initially have had a lot to do with such popular discharges of grief and grotesquery. But by the end of the Civil War the habit of whiteness had become so much a part of the social structure that there was no going back. White identity assumed the form of a hegemonic force within the social and cultural organization of political space in the country that admitted no easy alteration. It reproduced itself afresh in each new historical situation and gave vent to the violence of its fantasies on whoever got in its way.

Once it had emerged as a "commonsense" category, whiteness could then begin to disappear and operate in an increasingly covert fashion as the underpinning of racism. No longer embodied in the supremacist plantation economy as the obvious correlate to and justification for the institution of slavery, whiteness "morphed" into a more discrete role. With the reemergence of the southern planter class into a reconstituted form of dominance after securing the Hayes Compromise of 1877 (removing Federal troops and ending Reconstruction) and especially after ruthlessly crushing the Populist experiment of the 1890s in poor white and poor black cooperation, the South legalized whiteness as the "separate but equal" preserve of the older order. The 1896 Plessy vs. Ferguson Supreme Court decision was quickly mobilized in public space by way of Jim Crow legal enactments. In the North, however, the fundamental character of whiteness gradually became that of a silent presupposition, a taken for-granted norm that continued to operate under the guise of assimilationist "color-blindedness." Frankenberg explores in depth how this color- and power-evasive discourse has continued to organize privilege and power disparately. Indeed, this dissimulating operation of whiteness as racism is the real target of her work, given her claim that this paradigm has not been displaced by the 1960s emergence of race-cognizance but rather has hegemonically incorporated elements of the latter into its own continuing dominance (Frankenberg, 15).

Under the regime of liberal democratic individualism race privilege is primarily lived rather than seen (Frankenberg, 135). Since the 1920s advent of an ethnicity paradigm of analyzing social difference in academic discourse, discourses proscribing recognition of color or power differentials have underwritten material relations of *de facto* segregation. Frankenberg notes, for instance, the "horror" evoked in a white middle-class child when a relative uses racially explicit language (Frankenberg, 146). Racism, in this repertoire, is a thing to be overcome by acts of individual self-cleansing (Frankenberg, 168). Virtue is equivalent to establishing oneself as inhabiting a "noncolored" self—a goodness obtainable even by people of color to the degree their coloredness can be bracketed and ignored (Frankenberg, 147). Often enough, in middle class, "white bread" Americana, this proscription of racialized speech is reinforced by the isolation of segregated neighborhoods (Frankenberg, 240). Everyday experience anchors discursive denial. As we shall see below, it is this approach to dealing with race that the white suburb largely embodies and invites—despite the political developments since the 1960s that have pushed antiracist struggle into a different kind of battle.

Whiteness as Normativity

Even when the Civil Rights movement of the 1960s issued in a turn away from assimilationist integration to black political empowerment and black cultural conservation, white color-blindedness remained largely intact. Although multiple social movements for equality have insisted on explicit recognition of unique historical experiences of racial oppression suffered by Native American, Latin American, Asian American, and African American communities, whiteness has continued its career as a largely invisible norm of adjudication. Much of Frankenberg's ethnography confirms the difficulty of articulating—even among white women engaged in politically active antiracist efforts—the content of white identity and white culture. People "of color" are those with skins darker than white people, while the latter are thought to have no color at all. "Ethnicity" is not thought to include Anglo-Americanness, and even "white ethnics" are "ethnic" only with respect to their Italianness or their Jewishness or their Irishness, and not with respect to their whiteness.

The consequence is a subterfuge that masks how power and privilege continue to operate. Institutional organization of life chances remains constrained by subtle forms of racism that resist discursive clarification. Redlining in real estate, banking, and insurance practices continues to parse residential habitation, which largely determines quality of educational opportunity, which predestines employment possibility, which influences the likelihood of being arrested and incarcerated, which is now itself organized—in the form of the "prison-industrial complex"—as the new growth industry and largest employer in the country (Schlosser, 51–53).[1] "City" is the new euphemism for "race," and "young, male and urban" the new evidence of "criminality." "Diversity" is about "them," and "multiculturalism" is an initiative seeking to gain equality of attention with . . . what? The great unsaid here—organizing perception and privilege as that against which all else is measured—remains itself invisible and almost unthinkable.

Frankenberg details the peculiarity of the domination hiding in this latest incarnation of whiteness. Her basic argument is that contrary to awareness, whiteness itself is a cultural practice as well as an inheritance of privilege. It does not only designate a point of view but a form of embodiment that shapes experience. Its peculiarity today is its normativity. In comparison to the "bounded discrete spaces" of "other" cultures like those of blacks, Latinos, Native Americans, whiteness is experienced as generic and thin (Frankenberg, 192). Haltingly described by most of her interviewees in terms like "amorphous-ness," "mayonnaise," "blahness," "paleness," "neutrality," "emptiness," white identity is nonetheless not a form of cultural void (Frankenberg, 196, 199). It is the subject position of a profoundly powerful set of constituting practices rooted in domination.

The cultural specificity of whiteness, we could say, is precisely that culture whose content is not to "have culture" but to measure other cultures according to itself. It is not simply coterminous with racism or capitalism but is a way of being in the world that establishes itself in a silent contrast to others and yet does so in forms of practice that regularly borrow from those who are seen as "different." Exactly how such appropriations work has yet to be clarified theoretically, both with respect to their specific operations and their local histories (Frankenberg, 231–233). Part of the confusion that still surrounds white identity is the way it intersects with nationalism (Americanness), gender (maleness), class (the bourgeoise), and colonialism (transnational corporations) such that everyday life continues to "effect" or carry out forms of white domination over others. But what is clear is that whiteness remains profoundly relational even as it generates its own "individualistic" norms of being and forms of perceiving. Frankenberg is savvy in her priority. The task for now is not one of retrieving the "good" aspects of white heritage (Frankenberg, 232). It is rather that of unmasking its place-holding in the structure of power and liberating new possibilities for white antiracist participation with people of color in an ongoing struggle to transform the material relations of domination (Frankenberg, 233–234).

Strategic Postures[2]

If thoroughgoing analysis of explicit white practice demands a sustained recapitulation of the history of race, its implicit postures and presuppositions demand an equally thorough examination in the present. The relational aspect of whiteness noted above bears further elaboration in particular. While Frankenberg's exegesis offers helpful caution about speaking of whiteness as "thin" (and thus somehow "not cultural"), that adjective is perhaps not unwarranted as a designation in comparison with the blackness that is its discursive "opposite." As film critic Richard Dyer has noted, the specific character of whiteness as an analytical category today is its elusiveness (44). Its function as a norm means it is simultaneously everything and nothing, everywhere and nowhere. Its productivity is unlike the category of blackness, imposed by the dominant white culture as a pejorative stereotype, which has underwritten ever-renewed forms of white domination of the African American community and has been continuously re-fashioned by that latter community into ever-new strategies of resistance and survival (e.g., "black" power, "black" musics, "black" styles of speech). Whiteness has demanded very little by way of cultural creativity. Its ambiance is one of presupposed privilege, taken-for-granted access, expected protection, unhindered mobility, unthinking facility (Harris, 1; Haymes, 4). It has not come into being as a form of overcoming but rather as a form of plunder. It designates neither commonly shared suffering (as does blackness) nor the kind of communally initiated and tested artistry we find commonly expressed in African American individuals as a means of transforming such suffering.

Whiteness as a Predicament of (Un)Consciousness

It is thus not simply "lack" that shows up when white people "hem and haw" in uncertainty in response to a question about their ethnic identities; it is no surprise either that Nightline's 1996 production *America in Black and White* should turn up an assertion by white people from around the country that they are not obsessed with race "like black folks are." The muteness is part of the normativity: whiteness is "thick" (only) by way of unconscious habit not vigilant performance. It does not require thought or attention, neither the labor of articulation nor the deliberation of reinvention. It simply "is." Dyer even reports that for some white people, white racial identity is experienced as so ephemeral and void, it is spoken of in terms of "deadness" or even "death" when compared to (what seems to be) black vitality (Dyer, 44).

That death should emerge as a theme within the ideology of color-blindness is not surprising. Phenomenologically, it could hardly be otherwise, given the history of white violence toward black people. Death is one of the historical realities giving definition to the interactions between black and white in the United States. That history has produced the kind of depth-structure in the culture that Charles Long and bell hooks have theorized as the root experience of being black (as we have seen): the experience of living with terror, knowing that in the blink of an eye "death" could descend at any moment (Long, 1986, 158–171; hooks, 1992, 169–171). It could show up in the overt form of police brutalization (in our day the litany continues apace with the likes of Malice Green, Emmanuel Squires, Errol Shaw, and Dwight Turner in Detroit, Johnny Gammage in Pittsburgh, and Carlton Brown, Abner Louima, Amadou Diallo, and Patrick Dorismond in New York, and now Nathaniel Jones in Cincinnati). Or it could offer its erosions in the more covert forms of real estate redlining, mortgage and credit refusals, jacked up insurance rates, employment and promotion discriminations, mall surveillances, health care dismissals, restaurant waits, that slowly "dis-ease" the black body into an early grave. But the structure takes its toll not only on the black side of the line. What was and is given recurrent social expression toward blacks as a structure of violent oppression could not *not* have a reflex effect in white forms of subjectivity.

It is worth revisiting Hegel's phenomenological reprise (invoked in chapter 3) of the ironies that obtain when social identities get caught in a relationship of unremitting and hierarchical opposition (Hegel, 117–119; Fanon, 216–222). The dialectical reversal over the issue of death worked out by Hegel in his master/slave dynamic (in *Phenomenology*) throws black/white relations in our day in suggestive relief. By struggling with the ever-present threat of death at the hands of the master, Hegel says, the slave is shaken to the utter core by terror. "Being" itself is rendered insubstantial. But in being thus shivered in absolute fear, Hegel adds, the slave actually discovers the real power of *subjective consciousness*, its utter plasticity, its volatility, its ability to inflect even the direst circumstance with a modicum of meaning and value. On the other hand, in being forced to labor against the hardness of objective reality, the slave also crystalizes an *objective substance* of the self, an "otherness" yet carrying the stamp of the slave's own effort. In fact, for Hegel, it is the slaves who potentially gather a measure of autonomy to themselves, not the master. It is the slaves who constellate a certainty of independence in their consciousness, as a form of unhappiness. While that unhappiness itself marks the situation as one of injustice and demands redress, it also marks the slave as potentially more conscious than the master.

The master, on the other hand, is locked into an impossible contradiction, a fiction of independence and truth of dependence that admits no exit, because it offers no recognition. Having struggled with the "other" to the point of death and prevailed, the master then sets up the slave as a prophylactic of sorts, a security against the harshness of having to carve out an existence from an unsubmissive "natural" habitat. Neither does the master any longer face destiny and look life in the eye in the gaze of an unsubmissive other who demands equal respect . . . or fathom the death-threat that also shimmers there as a sign of the Ultimate Otherness that gives human being its deepest reference point. Instead, a kind of mistaken projection is set up, a kind of banishment of the function of fear inside the body; the pedagogy of fear is rather rejected and reorganized as "terror" against the enslaved. The master, we could perhaps say, unwittingly cultivates a hollowness on the inside of identity, a point of unexplored "darkness" that is maintained as an unconscious border somewhere between the deepest interior of the body and its lived surface. Fear is not fully faced in this body but instead takes up residence as a possessing other, striking out at whim. If the slave is unhappily dis-possessed, the master is unconsciously pre-possessed, lost in a labyrinth of fiction claimed as fact and bondage proclaimed as freedom. And it is at least ironic, if not quite revealing, that Hegel leaves this master dangling in indeterminacy in his dialectic, with no further consideration given to the possibilities for change. It is as if to say, for Hegel, there is no liberation for the master as master.

What Hegel highlights for the master/slave dialectic is suggestive for the dialectic of domination/subordination in the more contemporary key of color evasion. Whiteness today functions as a kind of silent *prophylaxis*, policing the borders between (its) more privileged lifeworlds and the social conditions it identifies as "black" and "dangerous." To the degree those social conditions (concentrated impoverishment, exposure to violent drug trades, proximity to toxic waste sites, etc.) in fact result in higher mortality rates and indeed constitute a "living form of social death," whiteness emerges as a structure of both avoidance and denial. In the same moment and (from a Hegelian perspective) "necessarily," it also emerges as a fiction denying access to life. There is a strange issue of disbelief that bedevils this particular structure of race. James Baldwin once put it this way: "White Americans do not believe in death, and this is why the darkness of my skin so intimidates them. And this is also why the presence of the Negro in this country can bring about its destruction" (Baldwin, 106).

The experience of whiteness *as* death (noted by Dyer) and white disbelief *in* death (noted by Baldwin) are flip sides of the same coin. Whiteness is not innocent of an "ultimate" inflection or "absolute" significance. To the degree white identity covertly mediates life chances, it simultaneously rations death encounters. By dissembling about the structural violence that underwrites its own well-policed privilege and accumulated advantage, it both ramifies and hides guilt. To be white in this world is not to be any less subject to death than being black. But it often does indicate a lifestyle committed to denying contingency and trying to banish the signs of mortality "elsewhere." But as Prozac use and substance abuse testify today, the ghosts come anyway. Theologically, whiteness constitutes a spiritual predicament. Neither the fact of death nor the demons of fear can be disinterred by mere projection—they have to be named and wrestled with. But it is not clear that the spirit of explicit white racism, once exorcised, does not return with seven more.[3] Naming and banishing is one thing; organizing a new identity is another (again as we shall see below).

Whiteness as a Dream of (the Middle) Class

But it is also not enough to characterize contemporary whiteness as a form of "social avoidance" or "death-denial" by way of Hegel's dialectic of *geist*. The color-and power-evasions also beg class analysis. The contemporary order of racialization reveals the dialectic of domination/subordination in its postmodern structure as profoundly spatialized. Increasingly, in both metropolitan developments of infrastructure and regional patterns of (im)migration, demography itself gives evidence of racialized organization—in the way in which race and class are made to ramify each other both materially and discursively.

Labor historian Mike Davis, for instance, in his writings on postriot Los Angeles, paints a provocative picture (Davis, 1992, 34). Tracing the events leading up to the eruption of South Central in 1992, he pieces together the local effects of national political decisions and transnational economic policies. Middle-class flight from the urban interior, job relocation, and ethnic turnover in the comprador class (in Los Angeles, the recent shift toward Korean control of small business in the area) find their structural correlates in the emergence of the suburb as the new center of the nation in terms both of votes and assets (Davis, 1993, 55). The advent of new monitoring technologies to guard entry and exit points of enclosed communities like Simi Valley is complemented by developments in satellite surveillance technologies that threaten, in the near future, to turn urban interiors into outdoor prisons (Dumm, 178).

The model here, according to Davis, is that of colonial occupation and control of racialized peripheries by the policing forces of metropolitan centers. Only the center is now the edge city and the periphery the urban core itself. Neighborhood spaces once constituted in "communal memory" are now reconstituted as "lifestyle enclaves," organized around the market values of individuality, materialism, and competition (Bellah, 71–75). The monitored norm of appearance, constituted in suburban consumption patterns, is, in fact, white, middle-class, heterosexual law-abidingness that (nonetheless) allows a certain range of commodified deviance (Dumm, 189). The surveilled "otherness," kept carefully and forcefully at bay, is presumed to be a "colored," lower-class, hypersexual loudness likely at any moment to erupt in a paroxysm of violence. Space is made to bear the new meanings of race.

In the middle of it all, discourse on the city euphemistically codes racist policy, while African American communities in particular (to a degree greater than other minority groups who also the suffer the effects) continue to bear the brunt of the "triage" practices perpetuated in the financial, insurance, and real estate industries. Banks and financial institutions function here as net drains of assets out of already impoverished communities and into wealthy "rim" areas. Meanwhile, the gap in services is filled in by a $300 billion "shadow banking" industry (finance companies, pawnshops, check cashing outlets, rent-to-own operations, etc.), legally charging (at times) as much as 240 percent interest per year, which supply huge profits to their behind-the-scenes Wall Street investors and financiers, Fortune 500 reputables like Ford, American Express, and various national banks (Hudson, 1–3). City neighborhoods that do become attractive to the middle class are themselves gentrified into predominantly white "pleasure spaces," enclosed by high-tech security systems, offering as allurement the possibility of consuming "black culture" (dance, music, sports, and fashion) shorn from its context of desperation (Haymes, 23).

On the other hand, given the constraints of residential containment, political abandonment, and economic disinvestment, gangs emerge as the "black market" enterprises mediating the drug dollars that alone keep eviscerated poor communities afloat. Since the replacement of the white heroin kingpins of the earlier part of the century by the Larry Hoovers of the crack cocaine commerce of the 1980s and 1990s, the death penalty has been re-inaugurated (Coleman, 216). At the same time, former American Association of Criminologists, president James Q. Wilson, develops the "somatotyping" profile of the kind of body that is statistically predisposed to commit crime (Dumm, 183). (According to what is already locked up, it is dark-colored, slim-waisted, heavily-muscled, broad-shouldered—the downside of pursuing the inner-city dream of exiting the 'hood by way of athletic prowess. In terms of statistical probability, the athletic body is also the criminal body.) By 1995 more than 30 percent of the African American male population between 20 and 30 years of age is under the "deviance management apparatus" of the judicial system. And so on.

Not far removed from this disheartening picture of the present, futurists like Jeremy Rifkin trace the historical logic of automation to predict a future in which not just minorities but the majority of people around the world will no longer be able to look to the commercial sector for a livelihood (Rifkin, xvii, 88). This futurist world is one in which the symbol manipulators and data-bank managers withdraw into their own enclosed enclaves while everybody else struggles to piece together an existence from temporary work, barter and service exchanges. In this vision historical and contemporary African American experience stands as the sign of the destiny of many others, including white people unable to secure overclass advantage for themselves. The apparition is one of life in an impoverished, largely jobless world of one-against-all opportunism and predatory lawlessness (Rifkin, 88).

In such a Dantesque vision of the new "cosmic" structure, race and class coordinate a this-worldly dream of "paradise" in middle-class minds, and a "that-worldly" nightmare of hell for those walled out. "Blackness"—in conversation in the foyer, by the pool, on the sundeck—takes its *real* meaning from the terrain of urban lower class-ness. And "whiteness" is a trimmed lawn, "green space," and a friendly gatekeeper.

On the other hand, class and race find some of their most poignant and potent entanglements in the actual encounter on the underside. John Hartigan's *Racial Situations: Class Predicaments of Whiteness in Detroit*, for instance, has initiated a miocropolitical analysis of white-ness as a quite local phenomenon of cultural *performance* that differs by neighborhood and class position even within the same city. Hartigan is careful to stipulate the ways the categories of "white" and "black" continue to organize resources, power and opportunity at the macropolitical level of the nation in accordance with the prerogatives of middle class oriented white racism. But he is particularly concerned to push beyond mere recognition of social construction of these categories to examine how they actually operate on the ground in quite varied local circumstances. His ethnography makes for fascinating reading of very subtle and sophisticated performances of such signifiers by (among others) long-term urbanite "hillbillies" who are themselves explicitly excluded from the meaning of whiteness as "middle class-ness" by their "white flight" relatives and former neighbors (Hartigan, 120–121).

In the racially mixed low-income neighborhood of Briggs near the old Tiger Stadium, for instance, the use and abuse of such terms is neither given nor entirely predictable (Hartigan, 120). Rather, Hartigan exhibits them as part of various locally mobilized "discursive reper-toires" that are regularly deployed in heuristic negotiations of unstable identities—creating humor, avoiding or instigating violence, probing power, erecting or challenging uncertain borders in bars and play-grounds. Far from fixed or certain, the words "white" and "black" on the streets of poverty are volatile brokers of invisible histories in con-crete encounters. They are regularly "played with," in ongoing attempts to anticipate danger, displace stigma, relativize the pain of middle-class rejection. Here, even "white trash" may, on occasion, proudly pimp itself as "just another N . . ." as Hartigan makes us aware (Hartigan, 114–115, 124–125).

But whether at the level of macrostructural social formation or the micropolitics of the street, these dimensions of race generally elude *middle-class* consideration. The latter's actual social location, demo-graphic mobility, and institutional access are not normally subjected to critical self-reflection, nor are racial categories made the objects of "playful" (heuristic) investigation. Space, we might say, shapes habit.

As I have detailed elsewhere, each of the two technologies of moni-toring and surveillance identified above has repercussions in terms of enculturation and ways of living the body (Perkinson, 2002, 184–185). The technologies of "super-vision" have historically consti-tuted blackness. As Du Bois argued in *Souls*, black identity first became palpable for him as the product of a particular mode of gaze, an imposition of the eye. The black community has *had* to become adept at displacing surveillance and invasion of its "room to be" by the white other (whether in the form of spying or supervising, rape or whip, baton or prison bar, social worker in the house or doctrine in the head). That such intrusion should go hi-tech today represents but a new wrinkle on an old problem. The community under duress has reacted by developing all manner of pedagogies for performance that deflect surveillance by proliferating the gazes before which the body-on-display multiplies its capacities to "hide in plain view." The result is a schooling of the black body—not always and everywhere certainly, but often enough and intensely—in an oppositional competency that rewards virtuosity in everyday "theaters in the round" (wherever black folks gather in barber shop or beauty salon, front porch or back stoop, basketball court, lunchroom or on the living room carpet).

The white middle-class body, on the other hand, is more often disciplined under behavioral norms more like monitoring, where the reward is for conformity, fitting in, not making a scene. This is a body that is generally "isomorphic" to its space—unschooled in using its surface as a site of contestation or subterfuge and unexercised in developing interior depth and "texture." Rather than growing up under intense forms of local community that act as a confirming chorus and supply a range of innovatory models and improvisational motives, white youngsters are more generally offered generic ideals and quiet affirmations. Not surprisingly, such a body has little experience of creating its identity under a hard eye or quick tongue—or if it does in a household of abuse, no chorus counsels comedy or invokes artistry as a requisite mode of survival. Where racial experience and communal practice make of the black body a battleground, enjoining performative competence to encode alterative meaning, not beholden to the eye that derides, white embodiment is rarely required to labor its interior into creative difference. It more often lives its "possession" (by the system of whiteness), unthinking.

Whiteness as a Cult(ure) of Desire

Cross-racial street savvy about racial categories (as noted above by Hartigan) not withstanding, race remains a broad field of social differentiation in America that continues to set up "black" and "white" as the extremes of material (if not thinkable) difference. In the country at large, they continue to organize the mutually exclusive poles of a hierarchy of otherness. Even where multicultural initiatives have forced some degree of "color cognizance," white identity, as we have seen, has remained largely exempt from examination or self-questioning. It has not remained innocent of its powers of projection, however. Social discourse increasingly recognizes the cultural particularities of experiences subsumed under generalizing categories like "Asian American," "Latin American," "Native American," "African American," and so on. (Although not surprisingly, the degree to which non-Euros are embraced within the ambit of "Americanness" by the dominant culture depends upon their context and their proximity to other groups of "others).") Occasionally, this cognizance even acknowledges the particularity of Anglo- or Euro-American experience—though as Frankenberg demonstrates, this does not immediately translate into ready articulation of such. Indeed, "American-ness" can be joined by hyphen to almost any ethnic designation. No one to date, however, is ever called "black-white." Parents of biracial children regularly describe the anguish of educating such children for the near inevitability of their being marked socially as "colored" and targeted for treatment in kind.

In the continuing hegemony of the older paradigm of assimilation, "blackness" continues to function as the necessary exception that leverages the paradigm. It anchors a schema of perception and a materiality of the everyday that requires what is unassimilable for its operation. Normative whiteness can be itself only by not being everything: it must have its incorrigible opposite. Blackness marks an absolute outside of whiteness that can never be "melted" away. As we have already noted, even dark-skinned people can "ascend the color gradient" into the social privileges organized as "whiteness" to the degree they are willing and able to distance themselves from the more pejorative ascriptions assigned to "blackness" that are associated with the inner city. By appropriating, for instance, signs of "safe middle class-ness" in terms of the kind of car driven, clothes worn, hair style chosen, English vernacular spoken, clubs frequented, and so on, "buppies" can gain uneasy admittance into gated suburbs (though such an address does nothing to insure against being "profiled" while driving, beaten while questioned, or even killed while being "checked out"). The racial schema itself, however, remains rooted at its bottom in a dense web of underclass associations (as "criminal," "bestial," "promiscuous," "violent," etc.) that constitute the "real" meaning of black skin in much of white imagination.

But these mutually exclusive poles of "white" and "black" are not merely references for a form of difference that is understood to be absolute in the dominant culture. They also mark out a gendered set of meanings that arguably enter into the ways white males and females are socialized and the way social space itself is organized.

Frankenberg notes in her research the frequency with which white women talked of their fear of blacks—especially black males (Frankenberg, 50). While careful to underscore that, in reality, this represents a mode of inversion—the history of race relations points overwhelmingly toward whiteness as "what must be feared" by blacks rather than the other way around—Frankenberg wants to avoid generalization. She calls for analysis and explores the discourses that attend interracial erotics. Her conclusions are that although white women involved in interracial intimacy (inevitably) negotiate their whiteness in highly individualized ways, they also almost inevitably run up against the way desire itself is racialized in our society. As revealed "in discourses directed against interracial relationships," constructions of white femininity are not innocent of constructions of racialized masculinity (Frankenberg, 237). In such discourses, white women and white men seem to be positioned, respectively, as "victim" and "rescuer" (Frankenberg, 237). She further recognizes that white girls often evince fear in relationship to peers of color and speculates whether white boys are not socialized into sentiments of hostility rather than fear in interracial situations. In either case, the idea that erotics is already racialized in the very construction of gender identities in our society is provocative.

While Frankenberg is careful to indicate the limits of her own study, her speculations are worth extending as question marks about the general social imaginary in the United States. To what degree would an archaeology of what bell hooks continues to describe as "white supremacist capitalist patriarchy" uncover, in its dominant middle-class assumptions, something like the following (stereo-)typology of both "figures" and "domains" at the turn of the millennium? (hooks and West, 160):

- The white male remains stereotypically located in the board room, "executing" power and privilege in decisions subject to review only by other white males. And even though most white males are not located "there" in fact, that space does remain the yardstick by which their other occupations, and at some subliminal level, their masculinity, are culturally evaluated and economically valued. (Indeed, it is the continuing hegemony of the "hero-conqueror" image of masculinity, developed in the frontier history of the country and given modern form in the nineteenth-century "captain of industry" mystique that now informs much of the ethos of CEO masculinity.)[4]
- The white female also remains, in myth and male mentality (not in statistical fact or real social experience), wrapped up in the cares of the bourgeois home, nurturing the fiction called the nuclear family, and if radically enough socialized to the role, perhaps even appearing at the door at 6:00 PM literally wrapped in a *Victoria's Secret* commercial. Again, even when the reality is nowhere near the fantasy, the image remains powerfully effective and affective in organizing both white male desire and white female psyche. And even when the white female is admitted into the corporate sphere, the male presumption remains very much alive that her role should continue to be largely that of domesticating, humanizing, and "surrounding in softness" what otherwise remains a "hardball" masculine realm of bottom-line "realism" and "takeover" ruthlessness (Fraser, 113–128).

- The black female is given place in this space of concretized myth as socially dependent and deviant, a "welfare Jezebel" or domestic "Aunt Jemima," whose womb is her only womanhood, the figural matriarch of a feminized household of color of three generations or more. As "labor," she can be admitted into the white male-controlled corporate space as ignored adjunct or exotic *ingenue* or into the private space still largely controlled by the white female as "surrogate mamma" without threatening the regime of desire in either domain (Fraser, 144–159).

- But the black male, finally, remains the wild card, the dangerous disrupter of the pleasurescape, the great grotesque "uncontainable," made both to signify and occupy the absolute "outside" of the structure as that against which the culture must defend itself—a virtual criminal-on-the-loose, sticking either penis or pistol into vulnerable flesh unless managed and monitored, in ghetto-block and cell-block, with ruthless certainty by the criminal justice system, or controlled, in the sports arena, with fictitious familiarity by the down-looking gaze and the dominating dollar of ownership. His presence is least tolerated and most suspect in the spaces of white male control and his "place" quickly (rein)forced at the furthest possible institutional remove from the privileged domain of macho self-realization. "Prison," in this compass, is the demographic opposite of "gated community" and the stadium, the benign equivalent of surveillance.

Obviously (!) the stereotypes are old and inaccurate. Would that their social structural correlates were equally as decrepit and bathetic! But *pathos* rather than *bathos* is the word for the way the myth continues to gather structure to itself. The myth encodes a desire that has a gender, an orientation, and a social location. It is male, heterosexual, and middle class—educated in a history of colonial takeover of bodies of color and their labor from around the world, realized in a globalizing enterprise that continues that presumptive right to resources and riches gathered from "equals" erased by the stereotypes. That the rituals of racial domination—as early as Irish insurgence against fellow black slum-dwellers in the mid-nineteenth century and as late as Appalachian lynch mobs of the mid-twentieth century—could not simply kill, but found it apposite to castrate, speaks loudly of the history. The darkness was a threat with an anatomical locus.

On the domestic scene today, everything from suburban infrastructure to condominium architecture, from federal budget percentage to local tax millage, from prison-industrial complex to gun-purchasing reflex, from mall security design to municipal loitering fine, from auto "Club" sales to auto cellphone details, is still decided with the (supposedly) predatory black male in full (though often denied) view. The desire that orders the governing perceptions focuses its fantasy on the white woman and its fear on the black man. Whatever anxieties obtain in the erotic space between white patriarchal attempts at coercion and control (of the white woman) and white feminist reactions and counter-assertions are haunted by a racialized specter.

The black male is made to do duty as both dark dangerousness and exotic eroticism—that against which the white male feels some deep need to defend his sexual interest and that by which he (perhaps unconsciously) measures his own sexual attractiveness. It is neither accidental nor ironic that prison construction is such a huge source of private profit in the United States and that rap now sells virtually every class of commodity (except perhaps Hammond organs) or that sneakers emerge as the fetish of choice. Black aggressivity remains an organizing anxiety and black "attitude" a coveted appropriation. The paradox is prodigious in our history and far from palliated. The seemingly inexhaustible phenomenon of a simultaneous white male fear of—and fascination with—stereotypic male blackness is as constitutive of "white" social controls in the present as it is of "white" identity in history.

Profiling Whiteness

In previous chapters, we have solicited scholars like Du Bois and Long, hooks and Gilroy, to clarify the degree to which terror has been the primordial experience of whiteness for black people and the displacement of terror through ritual work one of the primary modalities of black survival. In characterizing whiteness in this chapter, such fear necessarily marks a crucial concern. The Fear in question, the Terror troubling the land like an unwanted rain, like a cloud of toxin bursting from the factories of power and privilege, is neither one thing nor another. It moves godlike and inscrutable under the surface of American society, in the crannies of culture, ghosting streets, haunting handshakes, leaping out of eyeballs like a dart of envy or enmity, incarnate in economy and psyche, boardroom and bedroom, alike. And yet, it is also only a specter, nothing really, a mere reflex of history that is easily disavowed and denied. The encounter with racialized fear that has been a profound form of pedagogy in the black community meets largely with denial in the white.

Perhaps one of the most cogent and difficult things that must be said about white identity in the modern world is that it is fundamentally a structure of denial. It is elusive to talk about, as we have noted, precisely in its function of hiding history and domination under a presumed normalcy and a naturalized superiority. What has been hidden in particular is the degree to which white identity and white wherewithal socially, politically, economically, and culturally, are relational phenomena, gathered from a history of exploiting and oppressing peoples of color that remains largely opaque—ungrasped and unfelt—by white people. This lack of apprehension, this deafness and blindness to the cost of one's comforts and control for others elsewhere, chapter 5 attempted to throw in relief. But for most whites, the idea that whiteness is a primary meaning of terror for many black people is almost unthinkable. Ultimately, there is no remedy for such denial if black voices themselves are simply dismissed in their testimony to the reality they have lived and suffered. But there is provocative corroboration when we consider the reporting of a recent film like *Bowling for Colombine*, that witnesses to a level of fear in America that is astounding when compared with neighboring Canada, which is as much in love with guns and as much enamored with cinematic gore as its southern neighbor, but does not lock its doors, or bury its murdered citizens, with nearly the same frequency. The film exhibited a country virtually constituted in fear and hinted that the "reigning" subtext of terror—and its manifestations in murder and easy political manipulation—are rooted in an unresolved history of slavery and race. At some level, America is a terrain of terror for *all* of us, not just black people.

But it is a fear that is experienced and managed very differently depending upon one's powers of denial and avoidance, enfranchisement and geography. Charles Long's characterization of the dread experienced by William and Henry James framed the difference quite adroitly and with perspicacious relentlessness. America is animated by a level of fear that an *individual* body is incapable of "processing." Only a "community body," in touch with the harsh and unspeakable realities of the history, working in the modalities of both myth and ritual, begins to be able to transfigure the terror with anything like creativity. White practices of embodiment, however, typically fight shy of such ritual work and the consciousness it encodes. What might be called "white ways of being" (white walking, white talking, white gesturing, etc.), "speak" of a peculiar kind of American "possession" precisely in what they leave unsaid in posture and unexplored in the body. But the claim only comes clear when the silent separation between black and white is breeched.

At the inchoate boundary where the ghetto ends and the rest of the country begins, where an "Eight Mile Road" marks more than a mere change of blocks, where space is cut by an invisible line of race, gods lurk, demons bray, the numinous confuses the senses. Speech patterns suddenly betray more than mere individual intention, words harbor strange resonances of other presences, embarrassment rises or anger descends. Bodies become possessed with fear and desire, with excessive intensities "shouting" hidden themes. (I have had suburban students come to class visibly trembling and verbally stuttering from having driven into the near downtown neighborhood where the school is located for the first time ever in their lives.)

But only those communities whose life chances have depended upon becoming conscious and conversant with such clues and cues of "otherness" have developed the competence to decipher the surplus. Black communities have named the nemesis variously—"Blue-eyed Devil," "The Man," "Massa" (in the slavery era), "Babylon" (if speaking the lingo of "Dread"), "The System" (in the 60s), or "The Thing" (if one of Toni Morrison's women of funk). Du Bois has offered the word "Terrible," Charles Long, "*Tremendum*," bell hooks, "Terror." The key is not accuracy, but the recognition of something uncanny, a "more than" meets the eye. Whiteness, for non-whites, is an incubus, the underside of the Dream, a down-pressing Nightmare. Whiteness, for whites however, yet awaits its spiritual decipherment in a crucible that will crucify (white) presumption and set free the human being caught in its webs of fiction. White people are not forever doomed to live the meanings of their white possession. But exorcism will require education of the senses to the way they are habituated in pattern and presumption. Here, something like the Pauline language of Principalities and Powers can hint at the stakes in an archetypal—if also stereotypical—manner.

The Place of Supremacy

White experience of material space—in the suburb, the gated community, the university, the mall,—we could gloss as an experience of "gossamer," a kind of living and moving and having one's being inside a delicate halo that protects and buffers. The buffer is not mere illusion, mere ghost of dream. It is in part a function of police and immigration law, of access to inheritance and city hall, of free movement across the pavement and through the air without encountering interrogation. Certainly not all white people may experience such, but the pattern and history are there: white skin, by itself, will never in America as currently constituted, draw down on its head the baton of rejection *because it is white*. Space cooperates with whiteness; white people have "place." Where America is "lithic" for many people of color, a wall of hardness and denial, even for those who are upwardly mobile, for whites it is the very meaning of plasticity.

In response, in habit, white ways of inhabiting space split into an unimpeded fascination and an uneducated fear. In mythic parlance, we could perhaps say whiteness emerges within its own "space of being" as an incarnate form of *Fascinans*, the unthought and godlike presumption that its habitat exists "for" it, lies open to its desire, is plastic before its mobility, available to its appetite. Space for it is not obstacle to be overcome, but invitation to sample. It is a surface to be invested with confirming signs, a set of objects to be organized into comfort.[5] From this angle, white space emerges as a House of Sameness in which desire forages with at least the freedom of anticipation if not of realization.

But whiteness is also a "haunted" experience of spatialization. Its dark other, kept "in place" by the economies and policies of differentiation, hovers at the edge of imagination, as the very form of Fear itself. White freedom to occupy space is bounded by a "commonsense" line, beyond which lies only danger and deviance, constructed in sound bites, fetishized in cinematography. Whether the effect of that line— whatever actual neighborhoods or communities it demarcates—is more a quarantine of blackness or a prophylactic for whiteness is irresolvable. What is clear, however, is that such a managed blackness is constitutive for the spatialization of white identity as "irresistible force." It marks the point beyond which white desire chooses not to go, except as a policing power of invasion and containment.

In spite of itself, and irreducibly, then, whiteness births a Manichean *cosmos*.[6] It materializes a landscape of twin powers, intertwined principalities. The history it refuses to face as its own horror (as we shall see in chapter 7) is projected as the wilderness or underworld it dare not know. The very fascination it mobilizes is animated by a shadow that simultaneously dismays. At once *Fascinans* and Fear, whiteness concretizes a troubled uncertainty on the map of spatialized meaning.

Carried back far enough, white constitution and containment within this narcissistic sameness encodes also its relationship of genocide to the aboriginal inhabitants of the land and the ancestral memories such a landscape bore. The homogenization of space witnessed in monopoly capitalism builds on that earlier refusal and annihilation. At this level of meaning, the very plasticity of space presumed in white ways of living mobile and unimpeded simultaneously underscores a certain loss of the specificities of place. It points to the fact that whiteness has never quite been at home in this country, never fully submitted itself to the hard-won teachings of local people and local places, local wisdom and local particularity (except where it has "gone native" and learned from soil and indigenous people—as in the case of someone like the grandfather in *The Education of Little Tree*—and then it has ceased to be "white"). It has rather remained restless, a lonely frontier power overrunning the landscape with its prodigal energies of reconstruction and change, wrapping itself in technology, reconstituting itself as cyborg. Between it and the ecological materiality that could have reflected and deepened identity stands a whole forgotten history. Under its feet, interdicting immediate access to its "chosen ground," runs a lost river of murmuring native blood. As an ongoing refusal of limitation, recurrent white "frontierism" has functioned as the exact correlate of capitalist expansion: a common coin of identification that volatilizes every specific attachment to place or past. Its corresponding association of blackness with preternatural danger, bestial deviance, and primitive desire merely gives archetypal expression to a long history of deferred self-confrontation and ecological displacement. The metonymic "unsaid"—the whole chain of significance that is simultaneously evoked and lost sight of in these dislocations and expressions—is the un-mourned loss of connection to native people and natural habitat. White is what has buried green underneath red by means of black.

The Timing of Propriety

The discourse typical of whiteness likewise offers a profile. This writing began with an example of what I labeled "white dissimulation," a "rambling inarticulateness" of four white male students to a challenge about race thrown down by three black female colleagues. The incapacity of the former was indicative of a particular competence—and incompetence—related to a particular use of speech that might be caricatured as instrumental and routinized. Protocols here enjoin "exchange values" of words, getting quickly to the ideas they can "sell," rather than delighting in play and proliferation, exploration and exhibition of a richly articulated inner life. Anthropologist Thomas Kochman's work on "black and white styles in conflict" (the actual title of his book) analyzes such at length. Writing out of his experience of classroom struggles between his white and black students during the 1960s and 1970s, Kochman differentiates the cultural pattern and classroom style of what he claims is "white, mainstream, middle class, and Anglo-American" as compared to the distinctive aesthetic of black culture (Kochman, 13). While undoubtedly starkly modified in recent years by the adoption of hip-hop attitude by youth of all cultures, the profile Kochman identifies certainly would apply to the classroom scene I described above, and more generally continues to exercise normative influence in white middle-class work and family settings across the country.

In particular, Kochman labors to clarify a "mismatch" in modes of engaging classroom material (Kochman, 30). Black/white differences in public debate of issues are especially evident to him in the realm of what he calls "spiritual intensity." By this he means the observable behavioral difference that translates as "high-keyed, animated, interpersonal and confrontational" for blacks and "low-keyed, dispassionate, impersonal, and non-challenging" for whites (Kochman, 18). For Kochman, each characteristic mode is also class-correlated (Kochman, 14, 18). Black style here is that of "community blacks," who even when they attain a middle-class lifestyle tend to remain conversant with and comfortable in distinctively "black" patterns and perspectives rooted in the experience of segregation. White style, on the other hand, in this analysis remains predominantly middle class. Beneath the obvious and stereotypic surface differentiation lie formal and functional differences that regularly erupt in classroom conflict.

Especially telling, according to Kochman, is the difference in approach to argumentation. Affect and emotional intensity are factors in both persuasion and ventilation for blacks, according to Kochman. In the case of the former, dynamic opposition and contention that appears to whites as hostility is actually part of a cooperative mode of testing ideas (Kochman, 19–20). Persuasion here is a function not of quoted authority, but of truth values established in the crucible of argument that is passionately and performatively pursued one idea at a time (Kochman, 24, 34). Competition between individuals for floor time is heated, and the right to such is authorized by the quality of the performance, not by a predetermined sequence of turn-taking presided over by an authority (Kochman, 26, 28). A refusal to contend vigorously is regarded as intransigence or insincerity.

Whites, on the other hand, tend to pursue truth in the mode of cool detachment. Their relationship with ideas is that of spokesperson rather than advocate. Dispassion and impersonality are held to be the appropriate mode for truth-seeking; expert-citation and quotation are the venue for establishing authority; and careful turn-taking is the governing protocol for the manner of proceeding (Kochman, 21, 24, 27). Where "out-of-turn" black interventions that are nonetheless germane (to the content under debate) will be considered to be "on time" according to black protocols, white process valorizes only "one person at a time" (Kochman, 28–29). "Black" process-time is thus populated with multiple voices in contention around a single idea, while "white" time witnesses a segmented process in which a single voice is followed by another single voice and so on, until the argument is concluded. Serious, methodical, purposeful understatement determines white exchanges and disvalues any animation and affective intensity as hostile (Kochman, 31).

In interaction with each other, black and white differences of approach are immediately confusing. In net effect, where whites infer and experience a loss of self-control when confronted with black intensity, blacks often are simply giving expression to cultural protocols for successful social concourse. According to Kochman, blacks have much greater practice in managing anger and verbal hostility without losing control than whites do, and consequently much greater confidence in self-assertion and free self-expression (Kochman, 51). They are able to trust animated, vigorous modes of interaction as a "balancing" mechanism for conflict in which reciprocal exchanges of intensity harmonize energies without leading the participants to feel overwhelmed. On the other hand, where blacks sense dissimulation and a silent withholding of care and concern on the part of whites,[7] the latter often are simply hunkering down in habitual forms of repression and trying to keep from losing control. Problems occur in the classroom (and everywhere else), when white insistence on calm, rational, unemotional, logical process is put forward as a taken-for-granted norm, and not recognized as already culturally coded and specific to a particular social position. Blacks often perceive the demand for such as a political requirement—and submission to it as political defeat before the fact (Kochman, 40).

Kochman further notes the grammar of black call and response as establishing a particular cadence (Kochman, 110). The intervals between responses are close, demanding almost simultaneous expression. Being "on time" or "dropping in" or "dropping out" is as much a part of the meaning structure as the content of what is said. As with black musical sensibility, "feeling is engendered to the extent the rhythms [of speech] conflict with or exhibit the pulse [of conversation] without destroying it altogether" (Kochman, 111). Timing is not necessarily a matter of being right on the beat, but of having kept track of it, and of displaying it and playing with it, with a certain degree of artistry. In white speech, on the other hand, "the intervals between times when emotions are aroused and when they can appropriately be expressed are much further apart" (Kochman, 112). "Good taste" dictates restraint until an appropriate moment, and a concern for maintaining poise before the other, rather than a concern for the depth and quality of one's own feelings themselves. Part of Kochman's point, here, is to underscore a difference of *habitus*, of the habitual and ongoing calculation of action and anticipated response that shapes any given social interaction. White speech anticipates a low level of capacity to manage intensity and emphasizes control in expression; black speech knows a greater facility in working with expressive vitality and presumes a kindred capacity in the other (Kochman, 117).

Expressed archetypically, we might gloss such "white-speak" as spirit-possession by the Order of Monotony (literally, the sound of a monotone), a training of the tongue that does not want to speak in tongues, host ancestors, or lose consciousness of its drool. It is proper speech to accompany being in the proper place. It is periodized in well-regulated time and timing, wedded to segregations of meaning, and turn-taking in argument, and clarity of intentionality rooted in a thing called an "individual." It is confused by polyphony—by communal possession of the space-of-speaking with a multiplicity-of-incantation: crowds of signifiers parading carnivals of affect and tears. It has trouble saying "hi" to the homeless. Such a whiteness of the mouth is anti-incarnational and terrified of resurrection, a habit of speech that is utterly certain the past, the present, and the future are discrete, and matter devoid of spirit. Of course, I am speaking in caricature: there is no white *logos* common to everyone with pink skin. But there is a pattern and pedagogy, a discipline of the lip that matches the organization of space and the training of the body. Whiteness is a conspiracy of silence about history, carefully buried inside a body not yet fully born into society.

The Body of Silence

But the "white dissimulation" noted at the beginning of the above section is not just an affliction of the tongue. More precisely, it is a dis-ease of the body. The product and producer of white space and white speech is white embodiment. A spirituality of the racialized "white body" in America, if such could be written, would necessarily have to account for a reflex of uncertainty or even timidity in bodily habituation. Such a hesitation—born of the violence undergirding one's position—would index what might be called "the narrow body," a kind of modern inverse of Mikhail Bakhtin's "grotesque body" associated with the ribaldry and raucous exchanges of fluids and meaning, foods and dreaming, times and seasons of folk-pageantry and carnival-revelry worldwide (including Europe in the middle ages) (Bakhtin, 11, 27, 32, 84). Rather than a body lived as "unfinished," open to the world, not clearly bounded away from other bodies (of ancestors and progeny, animals and plants, earth and air, and the airs and odors and issues of orifices of all kinds), the modern body, in Bakhtin's genealogy, has increasingly found itself cast as complete and isolated, alone, fenced off from the world (Bakhtin, 27, 29, 53).[8] It has ceased to be understood or lived, as it was in medieval European folk culture, as a link to a much larger whole. Hollowness, rather than protuberance, is its sign.

This citation is further reinforced by Marx's notion that the social organization of industrial production had a reflexive effect in the body of the individual worker. The idea is suggestive that by sheer dint of repetition, the detail worker becomes "a mere fragment of his [or her] own physicality," a "crippled monstrosity," carrying the sign of manufacture in his or her flesh like the brand of a chosen people (Marx, 1967, 360). I am proposing a similar effect in the realm of the symbolic reproduction of white racial identity. Over time, through unthinking repetition, racialized perception and racist social organization become coded in the individual body. This coding takes place by way of a thousand small messages and disciplines. These are not explicitly white, but go by the name of "fashion," "propriety," "taste," "education," and "beauty". They are promulgated by advertising, engendered in the family, reinforced in the schools, recreated in social institutions, routinized in the military, and reaped as profit in the corporate board room. They operate by way of mimesis at the most fundamental levels of identity. And in contemporary America, they are anchored in a "great chain of appearing" that reflects the nineteenth century's "great chain of being." At the bottom is black, inner city, and ugly; at the top is white, CEO, and refined. The bottom may well be commodified and marketed in popular culture as "attitude" (selling everything from sneakers to fajitas), but at the top, the canons of corporate whiteness remain firmly entrenched.

Throughout the space of America, a particular body presides, invisible and ruthless (though often caught sight of in the State of the Union address). It normally makes its presence felt only subtly. Feminist writing has been prescient in uncovering the degree to which topography in this country is constrained by a particular eyeball: everything from fashion to football, male gesture in the bar and on the bus to female ritual before the mirror or after the meeting, answers to its demand. Rodney King's body on the pavement bore the wrath of this inchoate force only at a distance, mediated through other bodies, other disciplines of the flesh. Its realization in society is layered, complex—a hegemonic power that does not long tolerate serious challenge, though it does not often reveal its lair. Its lair is ultimately a domain that is straight up, strident, and well resourced. Blue-uniformed bodies beat King; behind them was a black-robed barrister; and beyond that august presence, a gray-suited bearer of the dollar. Such bodies—as all bodies—bear the marks of their spaces of being and business. They are not innocent of their histories or their powers but are living codes, walking hieroglyphs, sending signals through chains of command and demeanor.

The hegemonic body in America, that in some measure determines the meaning of embodiment for most other ways of living human physicality in this culture, is quintessentially male, ruling class, heterosexual, and white. It does not control; it is not like the king's body in old times, but is more insidious, incarnating in architecture and assumption, rhythm of talk and pace of walk, gesture and posture and habit and influencing with silent proscription, vague condescension—until called out by other forms of embodiment. The white, male, middle-class body is its icon though not its real comportment. The white, male, lower-class body is its buffer though not its real protection. Even black bodies, brown bodies, male, female, and cultured, can serve as its surrogate, exhibiting its quiet code. Even theologians can shout its demand, silently. Behind these other bodies, the ruling body disappears into the depths of the commercial, the golden door behind the market. But when it is called out—by suddenly meeting a gay body, a transgendered body, an uncooperative white female body, a lower-class black female body gazing back with hard eye and sharp tongue—it frequently stumbles in dissimulation (or may rage with fists flailing if given sufficient privacy). But it is the black lower-class male body that most precisely touches the nerve. If such a body should manage to invade sacred "white space" uninvited, gesticulating its own love of animation and polyphony, or otherwise refusing the codes of propriety and control, look out! Whiteness will likely descend like a god and bear its tooth. What will appear, angry and howling inside a clenched jaw, full of laws and bullets, pulling the strings of all the other puppets, is a Hungry Ghost.

7

White Passage and Black Pedagogy

Hey Man, there's real people down here.

—Ralph Ellison in reply to Irving Howe (quoted in Long, 1986, 193).

In the midst of the Los Angeles debacle of 1992, footage was shown on national TV in which a white-looking 30-something man carrying a bag of "stuff" across the parking lot of a looted store is accosted by a black man who is about his same age. The white man is grinning. The black man is visibly upset. The irony is palpable. The black man swats the bag out of the white guy's arms onto the ground. Solidarity in the struggle is not the same as partnership in the parking lot.

At face value the presumption operating here seemed to be that, for the white person, the moment was carnival, *marti gras* in May in South Central. Free-bees and festivity for the opportunity-minded and the daring! It seemed to offer the prospect of a brief recovery of aspects of the white self normally repressed for the sake of law and order, discipline, and decorum. It seemed to promise ghetto-danger celebrated as titillating spectacle and populist sport.

For blacks, however—although not immune to such spirits of "party-over-here"—the meaning was much denser, more laden with history, more sodden with the inescapable. In black experience, the ghetto is not a holiday venture, but a slow-motion noose and an every-day hell. Pilfering back some of the product that regularly out-prices the community was simple revenge. Profit-taking in reverse without remorse. It was the same game played in the boardroom minus the suits and the suites, the lobbies, the laws, and the legitimacy. The grin of black carnival thus also leveled a serious eye of business. It was a moment when "what goes round comes round," in fact, did briefly come around.

Patently, this was a racialized moment—even if race is not a real thing. The ticket to admission at the level of the street was indeed "color"—as the common badge of a variously suffered experience of oppression and exploitation. And the payback was a potlatch from below. But whiteness, whatever other forms of suffering the person carrying the bag might have been burdened by, has never been that kind of sign of the subjected and subordinated. By self-definition, whiteness has always claimed the caveat of being "no color at all." Commonness with those who are colored is the very antithesis of its design. And so the bag was swatted to the ground.

The surprise, the consternation, the hurt, even, in the white man's face, in that moment of rejection—a hurt inexpressible, unassignable to a mere individual, irreducible to any single explanation or under-standing—is the subject of this chapter. It marked a brief registration, even though only brief, of the meaning of whiteness in white flesh. It was a tiny flicker of horror.

It was also a teachable moment, however, indeed a moment of deliverance, were only the teaching and the exorcising community present. But whatever in fact happened to our white-man-surprised-by-his-whiteness in the days or weeks or years following that mini-rite of challenge, I want to underscore that white healing necessarily passes through a pedagogy of horror or it passes nowhere.

If white people would come to grips with the ongoing legacy of white supremacy in this country, there is no escaping the moment of facing the demon. There is no refusing the yawning void that opened suddenly between the white man and his black challenger in the park-ing lot that day in LA, no denying its dizzying depths or the dark numinosity that haunts its hidden labyrinth. Like Jacob wrestling the angel at midnight when he reenters the ghetto of Israel after a long absence, terrified to face the brother (Esau) whose blessing and birthright he had stolen and consumed so many years before, there is a compelling Fear, with a capital "F," that demands confronting here. The passage to racial maturity, to multicultural competency, to real-world democracy in a world of real differences, for white people lies through a rite of initiation.

But "what" initiates is finally uncanny. It is unnameable and untameable, hair-raising and hip-displacing before ever it can become a grace of the new and the tender. It does not yield to doctrine. It does not grant vision. Even as grace it remains grave and ungraspable. Only as a process, as an "exile and return," does it grant a new possibility. But if like Jacob, in the hour of reckoning, all false succor, all the familiar comforts of ancestry and assets, are summarily set aside and life itself is wagered in the struggle to know, it is possible for white people to pass through their "night of unknowing" into a new name and task, a new possibility of being. It is possible to come through the midnight horror of one's own racialized history with a new word for oneself. But it is a word that is speakable only after the fact and at the cost of a hip and with the offering of one's entire herd of assets. White identity can return to itself, at peace, only through a dark night of the soul, of fighting against flight. Its stigmata, its sign of having fought without reserve, will be a limp at dawn. Its truth will be its dislocation, the breakup, without any severance, of its strength. What follows seeks merely to mark the daytime features of that nighttime struggle and the fear and faith it occasions.

White Horror and Healing

This chapter explores the possibilities of a kind of "initiatory healing" for white people committed to struggle against the machinations of white supremacy. The process envisioned is far from winsome, but enjoins rather a trek "down" into the shadow-lands of Western history, into the unconscious, into the netherworld of terror and uncertainty that underwrites our modern experience of anxiety and claustrophobia. Modern fascination with the macabre in cinema, iron-ically, is one place the patina has been punctured and the underside opened to exposure. Film critic Richard Dyer's analysis of the "living dead" film-series directed by George Romero, for instance, offers commentary suggestive for the kind of self-confrontation such an initi-ation might entail. *Contra* more common stereotypes, the *Dead* films give us a political allegory in which black men emerge as "heroes" who use their bodies creatively to survive while whites end up in thrall to a kind of mindless "living death" (Dyer, 59). The picture is emblem-atic. At issue for the white subject of modernity is precisely the issue of embodiment. Dyer does not mince words in his description. In *Night of the Living Dead*, white characters lose control of their bodies while still alive and "come back in the monstrously uncontrolled form of zombie-ness" (Dyer, 63). In the process, Dyer says, "the hysterical boundedness of the white body is grotesquely transgressed as whites/zombies gouge out living white arms, pull out organs, munch at orifices"; similarly "white overinvestment in the brain is mercilessly undermined as brains spatter against the wall" (Dyer, 63). What is finally pilloried, in such a scenario, as "the spectre of white loss of control"—over their own bodies and over all the other bodies necessary for the exploitations of capitalism—is thus also ironically revealed as "the heart of whiteness" (Dyer, 63).

The point is exactly on point. Starkly revealed, white supremacy functions historically as a kind of globalized system of "zombiefication," forcing dark (and some poor white) bodies to the brink of the grave and then pulling them back into regimes of stupefying labor for Western (and largely white) interests. But as the discussion on Hegel in chapter 6 and on the living dead films above indicate—the question of "who is really who?" in the world of the zombie may not be what it seems at first sight. But in any case, it is not merely curious that the genre of film yielding this kind of analysis is horror. Joseph Conrad's *Heart of Darkness* offers a similar hint about the historical emergence of white self-consciousness in its representation of the European colonial project in Africa. At the end of the quest to civilize and save, submerged in the scene he had set out to transform, the colonial adventurer Kurtz speaks the word that rebounds back through the entire text. He says simply, "the horror, the horror." When one finds oneself at the savage end of a seemingly beneficent enterprise—despite, and indeed even because of, all good intentions—the conclusion can only be horror. If black experience of white culture must be epitomized as "terror," white discovery of the meaning of whiteness in that same dialectic must necessarily pass through the moment of self-horror. Anything less is dissimulation and self-avoidance.

The Release of the Grotesque

Chapter 6 argued that whiteness coalesces historically in this country as a means of avoiding terror—not only the obvious terror of "the dark other," but also all the less immediate terrors that hide behind that more obvious one: the terror of unrelenting labor against an ever-prodigal "nature" (e.g., both externally in the fields of cotton and internally in the jungles of the soul); the terror of nature itself as an indecipherable domain of animal wildness and vegetative wilderness and human willfulness; the immaterial terror of a material body open both to birth and death, genesis from the ancestral mother and disappearance back into the earth mother. In a modernity increasingly walling itself off from these more traditional matrices by technological innovation, putting everything up for sale in the commodity, the loss of "place" is patent. Materiality, animality, plant vitality, water, earth, the body itself are increasingly realized as the objects of technique and titillation and automation and cease to be given the subjective powers of familiarity and instruction and witness. The spirit-teachers (the spirit-guides, animal familiars, and psychotropic plants) of traditional communities become the quaint figures of ethnographic writings, the commercialized fetishes of New Age quests, or the scintillating fantasies of a popular imagination craving conspiracy and extra-terrestrial apparition. On every side, the signs of local knowledges are ripped from their indigenous locales of meaning and made to circulate for a price, offering identity at bargain rates. Blackness becomes merely the arch-sign of what is most feared and desired. And in the mix, by projecting such blackness, whiteness comes into being as a general refusal of darkness—and of all the things it is made to signify—as somehow part of oneself. To the degree these are refused, they are infused with a terror that cannot quite be known, but only enacted as violence.

Also implicit in the caricatures of chapter 6, however, is an argument that there is need somehow to re-initiate what Bakhtin called the "grotesque body" as that into which one is born and reborn. Such a body would embrace the whole of life, including birth and death, ancestors and issue, local ecologies of the past and global biosphere of the future. In the absence of such, dominant culture members are too often left as adults in a preadult body (an inverse kind of grotesque body). This body is one that is uneducated in its own vulnerability and permeability, split off from itself in horror and from its other in terror—split off, curiously, from both history and destiny, both its own ethnic communality and its multicultural commonality, its particular genesis and its shared hope—and instead curled in on itself in an anxious knot of narrow self-awareness and concern.

In shorthand formulation, I am arguing that the white body must be returned somehow to its history, white identity reincarnated in local community and global cosmology. Blackness can no longer be erected as a buffer against the demands of maturity, a screen against which to play out fear and fantasy, despair and desire—the quintessential sign for what is wrong and the (negative) surety for what is right about "America." As Fanon already asserted in mid-century, finally there is no blackness. Neither is there whiteness. There is only cultural difference and social contradiction and political conflict seeking easy legitimacy in the signs of biology. One urgent task of the new millennium is that of undoing that easiness by clarifying the cost of its continuing history and imagining the vibrancy of a harder-won integrity. And here we rejoin the more inchoate "theological-ness of mainstream whiteness" with its more self-conscious practice as white theology.

White (God) Talk

A recent American Academy of Religions panel attempted to address exactly this concern under the rubric "How can white theologians begin to talk about race?" The concern was laudable but the question misleading. The entire writing so far here has been one long labored shout: "White theologians (as indeed white people in general) already do talk about race!" Just ask a Black or Womanist or Mujerista or Chicano or Dalit theologian. Or simply a person of color. The problem is that most white theologians "talk race" without knowing the language they speak. The language is one not so much of tongue (as we have seen, the "white" pink tongue is often halt and stuttering, often tied, precisely when it is loosed!), as of torso. Torso and tempo and tone and temper. It is also one of what is under the toes and what the toes move through, unthinkingly. White theologians' toes move through doors so easily the tongue scarcely even tastes the difference between the airs. The door opens, the dean, the secretary, the clerk, the banker, the broker, the boss, smiles; the entry is made, no "airs" needed. Inside is outside; outside is in. They are the same. The air of the white room is conditioned without air conditioning. For the white, there is no "exterior" problem, no dilemma of the outside, of the skin splitting space and "speaking" (and suffering) in unwanted ways, as Du Bois talked about a century ago. (But then, *mutatis mutandis*, there is no inside either—and here is the beginning of the problem!) It is all available without obstacle, without change. It is all conditioned, like air; all cool—indeed, like "ice" says poet Imiri Baraka. The "Ice People" who do not think they came from "Sun People." The ones for whom the temperature is always set, the door open, the skin unprickled by "heat".

No, I say, the problem with white theological talk is that it is almost always about race without ever mentioning race. This is its burden. It is an untaught pedagogy. A problem of the ear, of the whole body "as" an ear, failing to hear its own cadence, its walk. Whiteness is a walk without a talk, a talk unconscious of its walk, a modern meaning of "talking head" verbosity, oblivious of its body. That is the problem. How can white theologians begin to talk about race? By ceasing to talk.

But you will complain: "Jim, you are talking in metaphors, in riddles. You speak poetry and license, without accountability, without discipline. A 'logy' without a *theos*. No God in sight. Mere bombast and tears, mere fears. This is surely not theology." But the plaint is already a presupposition.

There is no escaping the metaphor, the artifice, of the Word. We search for the ultimate through the proximate, through what it is not—inevitably. The meaning for which we hunger is mediated. Augustine was right in confession: "I was without and you were within and without I sought you" (Augustine, 254). But the "you" here, today—in this topic of the tropics of race, the tropes of white and black, of red, of yellow, of skin—is not just God. Or, perhaps it is. The "god" within the human . . . and its real character. And its real removal. Its exorcism.

This is the place to talk spirits and possession, principalities and the powers to unmask them, unman them. For they are, whatever else they are in this place called America, prodigiously "manly"—at least in their own projections (as we have seen). Just ask the police in South Central, LA, or Detroit, or Brooklyn, or Cincinnati, or . . . on campus. Just ask Officer Morris of Inglewood fame in July of 2002. A man's man, beating a woman's man, a momma's son, in cuffs, in chains, in a cage. It is a liturgy. A cast of characters. The longest running play on "Broadway," the archetypical script of the country, a clear destiny, the basic burden—bearing bodies through Middle Passages, buying commodities on plantations, beating savages, guarding against ravages, repressing uprisings and revolutions of clenched fists in the Caribbean and keefa-ed heads in the East and thug riots in the ghetto. Beating back the onrushing hordes of night. A man—against the jungle. White man, at the frontier of night. Protecting the light. This is the gospel of white. As silent and inexorable as an institution. Ramming plungers up anuses. Holding the handcuffed head and hitting it squarely, as fairly as the constitution allows! And you are perhaps irritated now, this is not theology, this is sheer ramble and cant, sheer propaganda. Not theology!

I say: Behold white theology. Behold the Man. Behold white talk about race in the key of fist, of frisk, and bombing. Behold the beginning of America. Or perhaps we need to hold, as Langston Hughes once implied, that "America" does not yet exist. But this bombing, burning, beating whiteness exists. It has already begun.

The Beginning

So having begun, how to start over again? That is the better question. Nicodemus's question (John 3: 1–15). How to crawl back through the history and up into the womb of the beginning and come out again, a second time, wet, sticky, needing everything, an unformed newborn—accompanied by one's whiteness as a late after-birth, "good" only for burial or the med lab. . . . Maybe this is the litmus test for a reformulated, a reinitiated, a re-evolved white identity: is it good for testing and then offering to the worms? That *is* its destiny. We should not be shocked, that is the destiny of all flesh. It is time whiteness became *mere* flesh, just like any other. It has existed as a denial of such. That has been its sole historical modality: it is the great exception, the "not like." The Supreme Absolution. The Ultimate Better-Than. Over against which, all else, every other form of flesh, whether human or animal, is inferior. For which, every other form of flesh is there simply to be used. Black flesh, red flesh, yellow flesh. The flesh of bull and buffalo, the flesh of the planet. Just "there," in the Providence of the Divine, to be used and labored and devoured by the Great White Tooth. And let us not be nit-picky here. All of our well-intentioned verbal professions of global solidarity and ecological responsibility and justice for the poor to the good, very, very few of those of us who confess the vocation "theologian" (or merely "academic") in America actually live a lifestyle that is not! literally! consuming the planet. Whatever we confess in the mirror, most of us here live and eat and shop and empty the garbage on the "white" side of the global divide of consumption (Mills, 44, 53, 138).

The Presence

The first lesson, of this necessary lessening of whiteness, is that it has existed so far in history effectively as a form of "Absolute." An absolute right to everything on the face of the planet. In practice—a god. Not *the* God. But the very modern form of god-in-the-flesh. The very epitome (scandalously!) of Christianity—incarnation caricatured—the Jesus-God of the blue eye and fair hair. Whiteness is first of all "theological." Before it moves its tongue. It is already in the eye like an idol. Or more rightly, like a demon. Malcolm was not far wrong. Whiteness is demonic. (Not "white" people, as he came to acknowledge, but whiteness itself, as a system, as a "talk" that is a "walk.") Whiteness is the demon *in* the eye of theology, what looks out *from* that eye, as it devours the world in its rapacious organizing gaze. (Here is perhaps the perceptual equivalent of the way language is theorized in gender studies as always already the regime of the Father. Modern "looking" is always already part of the White gaze.) And that is why to "do theology" as an immediate response to the question of how white theologians shall speak about race is already to misspeak. "Theology"—ever since there has been such a thing as a university discipline—has never *not* been about race. This is the damnableness of the demand.

White supremacy—the beginning and major meaning and incorrigible presupposition of all race discourse as it has emerged in modernity—has always operated as a hidden *a priori* that only infrequently reveals itself as such and speaks its own name. More normally it operates "as" a norm, as the hidden ground from which "talk" takes off, in modern Eurocentric evaluations of reality and divinity, operating through other things (like "theology"). There has never *not* been an "other of color" in surreptitious view, when modern Christianity has parted its lips to speak (Earl, 13; Pagden, 1–2; Perkinson, 1999, 439). We can abstract and pretend, we can cooperate with the exact social dynamic modern Christianity came into being serving and sanctifying, and presume to cogitate Christ "above" the concreteness of enculturated flesh that is his major historical coordinate according to the tradition that traces its advent to his. We can mystify ourselves, and others into imagining that white supremacy is "present" and potent only when explicitly identified as such, only when objectified as a clear political option easily repudiated as an aberrant self-understanding in camps that are liberal and progressive. But such an assumption is indeed mystification and the very meaning of ideology in the world order that is globalizing capital.

The first modern supremacy was Christian. White supremacy became its simplest shorthand. The system of global domination that has emerged in consequence has as its major social artifact a white male body that is "theological" in its deepest recesses of articulation (Perkinson, 2002, 176–177). Modern European (and "American") Christianity has never *not* spoken "race." It has ever been—and thus far remains—the *habitus* and "unconscious" of the global system of white power and privilege. Whiteness is, thus far in the structure of modernity, an order of "ultimacy" that supplies the basic code for resource flows and bullet holes, for war and plunder. For populations of color, white skin is a loud message, speaking volumes long before the lip opens. For white theologians, then, the question is perhaps better asked: what are the conditions for the possibility of *ceasing* to speak of "race"?

The Paradigm

Said another way, I am suggesting that white theologians "begin" like Dante—not at the beginning, but in the middle of life's way, *lost*. We cannot get up the slope of clear illumination and certain articulation except by way of going back and down. Our Virgil will necessarily be our own most potent "other"—not the pagan poet of antiquity "leading" the Christian back down into Christianity's own primordial depth-structure, but the preeminent and primal "pagan" of white American presumption, the black African who has most irresistibly—of all of those plunged by the practice of whiteness into the netherworld hell of modernity's underside—been made to know "America" *de profundis*. But formulating this necessary turn to blackness as a "solicitation for guidance" is itself already part of the problematic, the dilemma, of whiteness. Doing so would one more time—and now for spiritual succor in addition to the material and psychic "salvations" white America has always sought in black labor and loquaciousness—burden Africa with the role of saving America.

The kind of "Virgil" apposite to guide white men through the hell and purgatory of our history, I would argue, is one like John Brown of abolition infamy, who, in his hour, plunged all the way to the bottom of the struggle against supremacy. This shadow from the turgid past elucidates the real stakes: the battle to be engaged is not mere penwork, but puts bodies and destinies on the line. Dante, in his time, sought help for his trek from a forerunner figurer of his own lived exile—the poet (Virgil) of Aeneas's travail, breathing quiet wisdom in the wood of confusion, for one who was already "out of his political place." We, however—not yet exiled from the *Fiorenzas* of our own fascinations—need a liver, not a writer, of dislocation, who will figure the necessary passage in the flesh. John Brown is the apparition of American history whose "faint appearance" must be recovered from its long silence. But while insisting that a white man's feet trace the necessary tracks, it yet remains "black" wisdom that best knows the way. Thus Dante's journey and Brown's tread will be here completed by the visionary seeing (once again) of Charles Long. Obviously, the itinerary necessary for "white" education can be traced in multiple modalities—but the one I pursue here is Alighieri's allegory read by way of black augury made credible through the guerilla-ghost of the 1859 Harpers Ferry raid.

The Guides

Charles Long

Long's framework must first be briefly re-visited so that the Dantesque journey back and down—in order to move forward and up—will "flash" with maximum meaning. Among Long's contributions to the study of religion, as already noted, has been his notion that the experience of colonization itself, for the colonized, must necessarily be understood as "religious" (Long, 1986, 164, 190, 196–198). This is so because for those made to "undergo" the West, in the colonial aggression foisted upon the rest of the globe by Europe-on-the-move from 1492 on, Westernization entailed violation at the level of both body and cosmology (Long, 1986, 9, 110, 177). Not only were countless millions[1] of aboriginal inhabitants of the African, American, and Asian continents "eradicated" by conquest, forced labor, and disease but at the same time, their mythic "structures of explanation" of their worlds were ruptured irreparably (Long, 1986, 123, 193). The Western "myth of origin," however, remained seemingly intact—imposing its vision on indigenous practice—and through its own "civilizing" violence gradually reconfigured much of the indigenous horizon of meaning (Long, 1986, 137–139). "God," in such an egress, emerged as both "continuity" and "champion," at once self-consistent and triumphant, the very embodiment of a project of fascinated takeover and (seemingly) irrefutable rightness.

Indigenous cultures, on the other hand, had no choice but to work toward ritual embodiment and mythical figurement of an Absolute gone "dark" and dreadful (Long, 1986, 139, 165, 179, 196). They were plunged into encounter with a God of the Grave and the Ghost, a grotesquery of divinity that admitted no propitiation and no mitigation. And they alone, in this great earthquake of matter and meaning called modernity, relearned something of the depths of Life inside the terrors of death. Their mythic figurement of an original "containment" experienced in the hard schoolyard of "nature" (their precontact "myth of origin") was here recapitulated in colonial constraints and coercions. They learned hardness in double mode—and developed, in response, a "lithic consciousness," working out survival strategies up against the unyielding walls of plantation and reservation, factory and penitentiary, ghetto and *barrio* and all the mental categories of ideology and hegemony made to reinforce such shackles of the flesh (Long, 1986, 153–154, 178, 197).

In Long's construction, such cultures labor to penetrate behind the appearance of their oppressors as a proximate "epiphany of terror" in order to face into—and paradoxically be given new faces by— encounter with the real Terror of all living being, the Absolute itself in its inscrutable physiognomy as *Tremendum* (Long, 1986, 166–167). Forced to face their ultimate contingency, they have also, and thereby, worked out a profound intimacy with what lies beyond death. Ontological extremity has mothered cultural artistry and spiritual vitality. The dominant social order, by comparison, hosts a pallid spirituality and a trivial pursuit of meaning, a congress with the Ultimate known only as Fascinating and Hospitable. At least such is the house of mirrors we set up here, as we descend with Dante toward the dark side of the American Dream.

Dante Alighieri

Long's exegesis of the entire circumstance of late modernity as "postcolonial"—in which the ongoing interpellation of indigenous cultures and peoples into Western economic priorities backed by military prerogatives continues apace—supplies the coordinates for the trek. The pedagogy required, for white theologians to school their tongues in what their bodies already mark and militate for peoples of color, is serious pilgrimage to the underside of the modern, descent back down into the sacrificial matrix of "darkness" that gave birth to vaunted "republican pride," revisitation of the very body of the fallen "angel of light" whose real material effect is a ferocious wind of cold for the denizens of those worlds made to disappear inside its impact crater. Obviously here, I am splicing the model of the *Inferno* onto the model of the colonial, suggesting that the annihilation and "reservation" of natives and the incarceration and enghettoization of blacks map the terrain of the modern "hell" that must be traversed by any wanna-be talker about "Holy" Spirit in our age.

For those whose skins already "speak for" them in "the color-code of curse"—before they open their mouths—this underworld of modernity is unavoidable. Even when precariously perched on the upward edge of mobility or insecurely ensconced in gated domains of stolen dollars, populations of color have recurrent experience of the cold current wafting from the bottom of the bottom. Whether all the way inside their bones, or at some lighter level of knowledge associated with skin and hair, the *tremendum* has generally made its touch felt. At any moment that touch could become "technologized" and cut as deeply as police baton or bullet or bomb fragment—even if merely "accidental" as in the vestibules of Amadou Diallo or the weddings of Afghanistan. Indeed, the likelihood of such "accidental visitations" is itself indication of the more entrenched machinations of the white supremacy they enact and protect.

For those whose privilege is protected by the windings of whiteness, the work to be done must indeed cut "to the bone"—of both culture and psyche. Dante's descent provides a trope for an ongoing venture in archaeology. The earth to be dug and the relics to be recovered are both inside the body and under the history. (For now, poetry conjures the contours; chapter 8 offers more practical direction.) The basic polemic to be wrestled with is Dante's distress in discovering that "moving up" meant a long, circuitous detour *down* into the terrifying underside of his own history and identity. The self-confrontation was relentless: things were not what they seemed on the surface. Personalities encountered proleptically and anachronistically in the spiralling declination constantly surprised. Dante found compatriots as well as outlaws, figures loved as well as hated, in the walk through the archetypes and stereotypes of his own universe of meaning. Hell harbored a haunting threat—the exile could never be sure when the next disclosure might be of something or someone close to his own deep affections. The "other" without was always invisibly interwoven with *his own interior otherness*, invoked and not entirely repudiated in the horror of seeing such consequential suffering. But here we get ahead of the hike.

The slant on Dante's circumambulation proposed in this writing is that offered by Franco Masciandaro in his *Dante as Dramatist: The Myth of the Earthly Paradise and Tragic Vision in the Divine Comedy.* And already in the title alone we are pointed toward some of the necessary delineations. Masciandaro takes pains to point out that the superscript, *Commedia*, is a bit of posturing, securing for Dante the right to employ the vernacular mode in a self-chosen style of lowliness. But the actual structure and cadence is tragic—what our commentator calls the contradictory rhythms of *caesura* and failure (Masciandaro, xx). The choice is not minor. The effect is that of a "theatrics of reading"—an ever-growing concatenation of scenes that abruptly shift over into their opposites—that arouse feeling in a dialectics that finds no closure in happy logic, but only in painful transcendence. Quickly in the poem-*cum*-peregrination, Dante seeks to ascend out of the dark wood directly toward the rising sun only to be confronted by a leopard; his accounts of both primal forest and looming beast oscillate—initial deliriums of terror giving rise immediately to rushes of unfounded nostalgia for the ancient Edenic relations of peace (Masciandaro, xiv, 9). Neither will succeed in arresting experience in certainty, however.

But the issue is set. Do we accede to the dream of "earthly paradise" through ignorance and avoidance of evil? Or is the evil already present before we begin, horrifyingly "there," troubling the garden's repose even before our emergence into consciousness and choice? Masciandaro's opinion is clear. It is not denial of the tragic, but embrace of its suffering that grants knowledge. Neither is it enough merely to talk the walk; the real initiation must move continuously from tragic discourse to its (dramatic) experience. The way is through terror and its dis(-comfitting)contents. There is no rest herein, only constant failure, each scenario fracturing into the next, like a cubist painting offered as broken-edged itinerary of the life-way. Between the edges, what emerges is an "infinitesimally narrow space," the "antirhythmic interruption," according to a quote of Holderlin by Masciandaro, the "unique interval" at the apex (or better, "nadir") of alternating representations when nonrepresentation intrudes in the form of ineluctable silence and catastrophe, when the tragedy climaxes in a revelation of the (unwanted) truth of divinity (xvi, xxv). In another vein, Masciandaro will say, the reader-as-pilgrim must move from "representation" to "performance," by way of the poem—from *resonances* "dispersed on the different planes of our life in the world, [whose] repercussions invite us to give greater depth to our own existence," to the *reverberations* that take place when we speak the poem we have merely only heard until that moment (Masciandaro, xvi).

Here is rumination for white education: we white theologians are always in danger of being fooled by talk. The key is depth-experience and a new kind of speaking, born of shaking, in which the heretofore "unutterable takes form and is known, if only imperfectly and provisionally, within the *caesura* or empty moment" (Masciandaro, xvi). And what is that *caesura* or gap? Sooner or later in the crucible of interracial encounter, whites are faced with facing the tragedy of race in the nakedness of a dark face. What threatens to appear, if such an epiphany is not immediately banished by easy explanation, is "death," a mini-apocalypse of whiteness that alone can confer the blindness through which, pardoxically, we begin to see. This is the abyss that opens under our feet and inside our stomachs when (if) we forego the supremacy of certainty that has until now served as our modern "birthright" and prophylactic against the gnawing darkness of contingency. "White" skin is essentially a latent denial of death. But let go of whiteness in the encounter with people of color . . . and we step off an inner cliff! Who are we then? How shall we speak and behave? What if we offend? Will we be raped, killed, consumed? Or—God forbid—laughed at? (After all—with respect to those other possibilities, we do still "own" the police.) But it is too late. We have already offended (and raped and killed and consumed). The guilt is bigger than our own little ego, our well-wrapped body of fear. The tragedy has already occurred. America *is* tragedy. But the news has so far only reached the dead and darkened. Those of us still clothed in white continue to believe in innocence, in suburbs.

Masciandaro continues, in a footnote citing Fergusson: Even in the second act of the *Commedia*, the *Purgatorio*, though not a drama like the *Inferno* but an epic, the rhythm remains tragic

> both as a whole and in detail . . . Because Dante keeps his eye always upon the tragic moving of the psyche itself, his vision, like that of Sophocles, is not limited by any of the forms of thought whereby we seek to fix our experience—in which we are idolatrously expiring, like the coral animal in its shell. (Masciandaro, xxiii; Fergusson, 40)

The image is exactly the visage of whiteness—an idolatry, coiled in fixity, whose animal is dying within. The work to undo such a hardened carapace is a labor of rhythmic interruption. New content is not enough. We cannot merely read our way out of the tragedy. The devil is in the cadence, the grammar that organizes (Nietzsche was right!). The break with white presumption requires more than new words, indeed more than a new melody line. The shell has to be cracked, stripped. What might coral feel like, absent its condensate?

The route to be traveled by the Dantean chorus, says Masciandaro quoting Lacoue-Labarthe, involves a readerly movement through "disarticulation," through a rhythmic break with the logic of exchange and alternation (Masciandaro, xxiv). Dante's genius is not alone in imagery. He writes his own dismemberment in sequential structure as well. The percussive piling-up of scenarios encodes the necessary *crescendo*. (Indeed, "percussive disarticulation" might well designate the kind of ongoing experience whiteness must undergo, in the theater of racial encounter, to become, finally, conscious of its own stifling limitations.) But again, it is a climb in the direction of below.

The course winds down—into the dark interior of social, political, spiritual topographies. At the bottom is a body of revulsion (Satan) whose hoary grotesquery must be literally embraced and traversed, not simply by the mind, but in sense and flesh. Only at the utter extremity of depravity—going hand over hand down the "Dark" One's midsection—will directions simply be reversed and the way down, become the way up. Until then, the syncopated structure of debriding revelations continues without interruption. It is worth examining briefly this sequential "induction into terror" that Dante's itinerary poses.

Mysterium Tremendum

We have already indicated the nature of the discomfiture. It is nature itself—unmediated confrontation with wild beasts and wild space, what history of religions scholars would readily identify as the originary mode of encounter that yields human experience of the divine. Eliade calls such *hierophany*—the sudden comprehension of the sacred precisely through the profane. Often enough this morphing of stone or tree, forest or four-legged creature into an apparition of divinity is uncanny, an occurrence at the edge of experience, in moments of "extremity." In this manner, as in the overall structure of the journey he recounts, Dante's "comedy" can be well comprehended as a form of what I would call "Christian shamanic proclivity"—a descent to the underworld to face the demon and fight temptation, the struggle up the formidable mountain of self-confrontation to secure the lost soul, a flight to "the blessed isles" to win vision. In such a shamanistic travail, what must be won often enough first appears as "monstrous" before revealing its "good." Beauty has to be wrestled from what seems beastly—both without and within. And the pedagogue of greatest power here is fear itself. Only when the human looks into the abyss of mortality, comes close—through fasting or pain or loss in the forest of unfamiliarity—to death, is spiritual insight and psychic regeneration obtained. Dante's trek conforms to type. Indeed, especially significant, in this account is Dante's tracing of the way his fear is rendered "absolute" before being succeeded by faith. For Masciandaro, the drama is thrown in relief by way of Hegel's understanding of the role of absolute fear (in the master–slave dialectic, as we have already seen) in educating to truth.

Masciandaro reads Dante's itinerary as a pedagogy of (Hegelian) fear, mediated through the monstrous and the wild. At each step, the Florentine tries to retain some semblance of control, some bit of "mastery" masquerading as "cleverness," but with finally vain effect. It is from fear alone, argues Masciandaro by way of an Allen Mandelbaum quote, that the "poet can gather ultimate energy . . . the absolute fear—of death" (Mandelbaum, xviii). Indeed, such a state is where Dante has started—as many indigenous communities would concur that any initiation process or vision quest must begin—setting out on his pilgrimage in a state "as *like* to death as one can get while still alive"(Mandelbaum, xviii). Dante himself describes it as a wood "so bitter—death is hardly more severe" (Alighieri, I, 7). But the actual "encounter" with contingency happens only in momentary episodes, repeated like a refrain that has its effect only to the degree it retains its surprise and so resists the pilgrim's reflex to tame its terror by talk. The sequences of surprise concatenate like a drum beat, throbbing inexorably, again and again, again and again catching the poet off-guard, through some new apparition of threat, whether bestial or human. The overall "teaching" is a "stripping"—of defenses, of domesticated versions of divinity, of insistence that the Spirit be sweet and not a Hound of Tooth. Dante is only slowly and under duress forced to face the "tremendous" side of God.

Or described more theologically: Dante is delivered through delirium to a basic datum: the *Trinity is first of all Tremendum*. The absolutizing of fear in Dante's experience is the analogue of the Ultimacy of Terror that Otto posited and Long extended. The refrain bears revisiting. Dante is doubly schooled in the trauma he resists. As one of the urban *literati*, he has lived at a comfortable remove from the first pedagogy of mortality through fang and talon, storm and stream-in-flood. Long's unpacking of myths of origin makes clear that these natural powers are the initial "initiation," the first manifes-tation of untameable divinity to human beings in the key of ecology, the hunter-gatherer knowledge that life is fragile before the forces of other lives and appetites, other laws of wilderness and wildness. But this is a key only those remaining nestled in a natural niche know about. Dante is, however, a creature of the "machinery" of city, buffered from the more radical rawness of natural life by walls and hearths, lighting and linen, and all the other inventions by which the political species veils itself (until the grave) from the biting tooth. But Dante *has*, on the other hand, been exiled from his city, and (at least in spiritual imagination if not in lifestyle) his "salvific passage through terror" does begin with the hard edge of nature, before his descent into the pit of history to encounter the other kind of hardness, produced by human *hubris* and envy. What Dante as indi-vidual is made to pass through as "spiritual" itinerary, however, whole cultures have been forced to endure as material circumstance in modernity.

Long's innovation is to have reimagined the pedagogy of fear in the circumstance of oppression. Here, history itself becomes initiatory for those made to undergo its travail and losses. Colonized peoples, by sheer dint of domination and the arts of resistance they have crafted to survive such, know divinity in a modality that domination is precisely committed to denying. It is the *Tremendum*, with all that its manifest epiphany has to teach about contingency and gratitude, mortality and compassion, that such sufferers "of" Europe have managed to embody in their ritual memory and politicized mythology. They know, at the level of the bone, that tragedy is irreducible and death real—and that life and meaning are projects "in spite of."

And it is no secret that it is precisely *this* knowing that domination itself seeks to sequester and quarantine among the conquered, as the very antithesis to its own projected endurance through time. At heart this is the ontological relationship between "1st World" and "3rd World," gated community and ghetto, white skin and color. Whiteness is finally an impossible dream—a fictive denial of death, a prophylactic forever bursting in the moment of intercourse, a fatuous hope that God is merely Fascination. Dante tracks the necessary itinerary for growing up spiritually. The way passes through tragedy and terror. Anything less is simply more of the sickening same. But in modernity—in contrast to the high Middle Ages—the rite of initiation passes profoundly through a peculiar history. Salvation *from white supremacy*, for those of us so encoded and comforted, necessarily means descent into the underworld of modernity there to discover the body of terror is ultimately one's own.

John Brown

But there is a last piece to be supplied the paradigm. As noted above, it is John Brown, on the landscape of American whiteness, whose tracks trace the way—even if we lament his final choice for violence (as I do). This is a Virgil with a difference—as a doer rather than designator, guiding not through the pointing, writing finger, but by his feet. He becomes a Virgil for us by first undergoing a Dantean descent for himself. What Brown ferreted out and figured in his day, must be re-comprehended and communicated in our own. In a word, his ideal was a simple seal upon the aspirations of the republican experiment. He actually believed that "to remember them that are in bounds, as bound with them" was the heart of freedom, much less theology. And he interfered accordingly. He died conscious only of having acted to right a grave wrong, indeed the "sum of all villainies" (Du Bois, 1997, 182).

That it is (once again) Du Bois who will here give us our version of Brown's burden and witness is simple justice. Du Bois himself merely follows on the heels of others offering accounts more precise and labored, but none so felicitous or favorable to "the little known but vastly important [view-point of the] inner development of the Negro" (Du Bois, 1997, xxv). Du Bois considered his book on Brown the favorite of his productions. He considered the life so extolled there to be messianic—on the order of revolutionary leaders like Toussaint L'Ouverture, Gabriel Prosser, and Nat Turner—and the real bellwether of the nineteenth century even as Washington's was of his era (Du Bois, 1997, xii, 1, 174–75). But Du Bois's labor was to speak the meaning of Brown's burning signature to his own (later) age. It is a work unfinished, or we would not still be here today, in the twenty-first century, asking the question that anchors this writing. Indeed, underscored as it hereby is in the penumbra of September 11, 2001, Brown's boldness and stark clarity can scarcely be shelved as requisite only for a ruder time of harsher battles. The blood of this issue has never ceased to flow for over half a millennium. It is the shame of white skin that most of us so closeted have lived and believed as if no blood has ever flowed to preserve our pallor and privilege. But Brown's considered judgment that "the crimes of this guilty land will never be purged away *but* with blood" did *not* find its *denouement* merely in the War of Uncivil Brothers that ended in 1865. It remains a disturbing indictment of a system that has never desisted from its organization of life and death by way of a globalizing line of color. The measuring calculus of that dark incubus remains lodged deep within the body of white maleness.

But we do well to trace the Brownian motion for a moment, by way of the Dantean drama, before returning to a postmodern mobilization of the meaning. Brown began Puritan, passionate, and poor (Du Bois, 1997, 2, 6, 8). Early wanderings in the wilds, struggles with a fear of Indians before discovering them friends, loss of a well-loved lamb and then of his own mother, quickly impressed the tragic-comedic rhythm of life on the young rambler. An adolescent encounter with white brutality toward a slave friend portended future preoccupations. But Brown went through his first half-life, somnambulant and silent, attending business, securing family, marked by growing solemnity as grief and loss marked his soul. Du Bois will name that soul-effect the entry within Brown of an "iron of bitterness"—the death of a son of four in 1831, his first wife and infant baby in 1832, four more children in 1843, and three more between 1846 and 1852. His brooding turn of mind would ever wrestle with the mystery of such agonies; never sure of calling, never satisfied with the veil that kept things dark, never fully at rest (Du Bois, 1997, 17). But it took failure at yet another level to bring Brown to the course that would stamp his personal history on that of his country like a perpetual question mark.

Initial work in the tannery business, and investment of surplus in banking went the way of prosperity in general in 1837, when presidential tinkering with the credit system sent the whole house of cards tumbling to the ground in a national panic (Du Bois, 1997, 20–26). Brown floundered then for some years in race-horse breeding and farming, until sheep-herding presented itself as his remedy for both bread-money and leisure time to study toward his growing conviction of a deeper calling. Here, both personal morality and national perfidy coalesced to once more plunge him into crisis. Committed to a sense of fair play and prices, organizing wool-growers to sell equitably to manufacturers, refusing to "corner" the market or combat unscrupulous practices with their like, he was finally, in 1842, forced by the industry titans into bankruptcy. But his economic experience only sharpened his racial analysis: a country in which "the majority have made gold their only hope" can also expect no hope in confronting their major economic shibboleth with moral suasion alone (Du Bois, 1997, 28). Brown began to brood more seriously on the deeper question of slavery.

He had not been idle in the interim, but only unfocused. An 1837 incident in the Congregational Church of Franklin, Ohio gave evidence of the slow boil that was rising in Brown's soul. Free blacks and fugitives were only marginally seated in the revival meeting then taking place in that Church and Brown took upon himself to call attention to the discrimination publicly and give up his family's place to the misplaced blacks. In time, the act received its censure (in the form of a letter) and one of Brown's sons records that the father "turned white with anger" (perhaps the only just mode of that particular color's social meaning) at this clear show of "pro-slavery diabolism" (and in that act, and its emotional aftermath, the son's own "theological shackles were a good deal broken" for good) (Du Bois, 1997, 40–41). In 1839, the tension came to a head. In response to the persecution stories of a black preacher visiting Brown at his home, the middle-aged man rose and bound himself and his family to unremitting war on slavery in a solemn and secret pact and then fell to his knees in imploration of mercy and blessing. The conversion did not translate into anything like exaltation, but only began the journey back down the path into the country's deepest tragedy. Du Bois uses terms like "slough of despond" and "labyrinth of disaster" to describe the Brownian plunge; it was (significantly for the author so speaking) the "dropping of a somber veil of fate"—very like the veil of more famous invocation when Du Bois earlier in his career described his own boyhood experience of the first taste of racial rejection (Du Bois, 1997, 23, 43; 1961, 17). Brown was inducted into "life behind the veil"—his own peculiar "white" positioning, to be sure, not the same as the veil ghosting black life, but a mysterious, unyielding veil nonetheless.

But there is more in the Du Boisian account that reminds of Dante as well. This seminal moment of turning to descend back down into what was loathed and feared is formulated by Brown as a peculiar "conversion," a turning from sin toward vocation (Du Bois, 1997, 44). Brown had until then been majoring in minors and minoring in what was to become central to his life's purpose. And he understood it as selfishness. God would heretofore attend to the nurture of his family. He would attend to the cries of the motherless and fatherless who were so rendered by their bondage as someone else's property. He thereafter became "the man of one idea, and that idea the extinction of slavery in the United States" (Du Bois, 1997, 59). That idea organized the rest of his life of resistance—for both himself and his family—in refusing to abide by the Fugitive Slave Act of 1850, and counseling the same for others, in freeing slaves by stealth in 1851 or threat of force later on, in striking the blow of blood-sacrifice on the plains of Kansas that forced the forces of antislavery to face the failure of tepid politics, in organizing finances and hatching the plan (the raid on the armory at Harpers Ferry) whose failure would finally succeed in sending the country into ultimate confrontation with itself and its destiny.

But Du Bois is savvy in keeping his account of this solo flight into the stark savagery undergirding American vanity contextualized and suggestive. He keeps in sharp view the role of topography as well as society. Brown was a man schooled in his sobriety by the mountains that came to occupy so significant a place in his scheme of slave revolt and redemption. "On the side of the Alleghanies he tended his sheep and dreamed his terrible dream," says Du Bois, and "it was the mystic, awful voice of the mountains that lured him to liberty, death and martyrdom within their wildest fastness" (Du Bois, 1997, 19). Like Dante in the face of forest and beast, learning social fictions and spiritual facts under tutelage to terror in "natural" form, Brown must be taught

his future vocation among his own, by confrontation and conforma-
tion in the wild. But that pedagogy among the peaks and valleys does
succeed in orienting his ear and soul to another serenade, a sadder
song rising from spilled blood and severed genealogy. By the decade of
the 1850s, Du Bois describes,

> A great unrest was on the land. It was not merely moral leadership from
> above—it was the push of physical and mental pain from beneath;—not
> simply the cry of the Abolitionist but the upstretching of the slave. The
> vision of the damned was stirring the western world and stirring black
> men as well as white. Something was forcing the issue—call it what you
> will, the Spirit of God or the spell of Africa. It came like some great
> grinding ground swell,—vast, indefinite, immeasurable but mighty, like
> the dark low whispering of some infinite disembodied voice—a riddle of
> the Sphinx. It tore men's souls and wrecked their faith. Women cried out
> as cried once that tall black sibyl, Sojourner Truth: "Frederick, is God
> dead?"
>
> "No," thundered the Douglas, towering above his Salem audience,
> "No, and because God is not dead, slavery can only end in blood." (Du
> Bois, 1997, 56)

And here is a last figuration. From 1847 on, according to Du Bois,
Frederick Douglas was Brown's chief black confidante and "in his
house Brown's Eastern campaign was started and largely carried on"
(Du Bois, 1997, 120). Yet, Douglas, as taken with Brown as he was,
could not finally enter his plan. Du Bois offers: "as with Douglas, so it
was practically with the Negro race. They believed in John Brown but
not in his plan. He touched their warm loving hearts but not their hard
heads" (Du Bois, 1997, 175–176). For those whose "whole life was
already a sacrifice," the plan of a gradual escalation of guerilla raids
(to free slaves) moving ever-deeper into the great Black Belt running
south from Harpers Ferry and rousing ever sharper response until the
issue was finally engaged once for all, appeared certain of failure and
failure was sure to fall on black heads most heavily. In large part
blacks declined to sacrifice yet more, though a few did throw in their
large hearts and resolute wills anyway. But the issue for Brown was
not one merely of listening to black counsels alongside black pain. It
was a matter of deciding, Bonhoeffer-like, when responsibility had
become unavoidable, and opting to act despite the odds, alone, if necessary, giving careful planning toward the possibility of success, but
finally considering the gesture as requisite even if the outcome was
tragedy. Brown did not merely look into the looking glass of tragic
national history; he embraced the vision as lifestyle.

In retrospect, Brown's life would take its measure from a small gesture, compelling, in its failure, a large effect (a number of historians would say Harpers Ferry was the real beginning of the Civil War). But that cast of "things attempted" would not come until late; in daily life, Brown accepted ongoing struggle as the price to be paid in clarifying for the country that "the cost of liberty is less than the price of repression" (Du Bois, 1997, xviii). He once calculated that the losses for those serving the free state cause in Kansas could be put at 120 days for 500 such men—together with stacks of grain wasted and burning—adding up to more than $150,000 (102–103). Brown himself used part of his small inheritance, never touched for the benefit of his family, to fund "the Service" and considered that his "line of duty" demanded such, "though it should destroy him and his family." And he once wrote about himself (sounding very much like a spiritual struggler from an earlier era),[2] in the midst of his Kansas initiative,

> And he leaves the states with the deepest sadness, that after exhausting his own small means, and with his family and with his brave men suffering hunger, cold, nakedness, and some of them sickness, wounds, imprisonment in irons with extreme cruel treatment, and others death; that, lying on the ground for months in the most sickly, unwholesome, and uncomfortable places, some of the time with the sick and wounded, destitute of shelter, hunted like wolves, and sustained in part by Indians; that after all this, in order to sustain a cause which every citizen of this "glorious Republic" is under equal moral obligation to do, and for the neglect of which he will be held accountable by God;—a cause in which every man, woman, and child of the entire human family has a deep and awful interest,—that when no wages are asked or expected, he cannot secure, amid all wealth, luxury, and extravagance of this "heaven-exalted" people, even the necessary supplies of the common soldier. "How the mighty are fallen!" (Du Bois, 1997, 107–108)

Brown's hell was on earth. But whence then the courage so to live— "for twenty years . . . to permit nothing to be in the way of my duty, neither wife, children, nor worldly goods?" (Du Bois, 1997, 92). In part, according to Du Bois, it was his conviction of "opportunity." In March of his last fateful year, Brown wrote to a friend,

> Certainly the cause is enough to live for, if not to—for. I have only had this one opportunity, in a life of nearly sixty years; and could I be continued ten times as long again, I might not have another equal opportunity. God has honored but comparatively a very small part of mankind with any possible chance for such mighty and soul-satisfying rewards. (Du Bois, 1997, 115–116)

Partly also, it was necessity. James M. Jones later wrote of the convention of leading Abolitionists and blacks called to ratify a set of rules for governing "a band of isolated people fighting for liberty," that Brown, in personal conversation with him,

> led me to think that he intended to sacrifice himself and a few of his followers for the purpose of arousing the people of the North from the stupor they were in on this subject. He seemed to think such sacrifice necessary to awaken the people from the deep sleep that had settled upon the minds of the whites of the North. He well knew that the sacrifice of any number of Negroes would have no effect. What he intended to do, so far as I could gather from his conversation, from time to time, was to emulate Arnold Winkelreid, the Swiss chieftain, when he threw himself upon the Austrian spearmen, crying, "Make way for Liberty."
> (Du Bois, 1997, 130–131)

But partly, also, it was Brown's familiarity with the "vision of the damned"—the "great grinding ground swell of the Spirit," or "spell of Africa" as Du Bois had otherwise described it, which "was stirring the entire western world," from beneath (above). Brown knew well the history of Abolition, the efforts of fugitive and free blacks over generations to discover instruments of struggle (societies, newspapers, conventions, Underground Railways, etc.), the revolts and rampages of Vessey and Turner, the oratory of Douglas, the organizing of Delaney. He had studied carefully "the wars in Haiti and the islands round about" (Du Bois, 1997, 107). "To most Americans," says Du Bois, "the inner striving of the Negro was a veiled and an unknown tale; they had heard of Douglas, they knew of fugitive slaves, but of the living, organized, struggling group that made both of these phenomena possible they had no conception" (Du Bois, 1997, 123). But John Brown knew this writhing history intimately—in the stories of those he invited to his home and visited in theirs, in the advice he both gave to and sought from them, in his capacity to feel, "as few white Americans have felt, the bitter tragedy of their lot." Du Bois's book celebrates as well as delineates, one "who of all Americans has perhaps come nearest to touching the real souls of black folk" (Du Bois, 1997, xxv). (The comparative mode here of "nearest"—in its double sense of both empathy and impossibility—is critical, as we shall see.) It is no surprise then, but rather testament to that propinquity of purpose and intimacy of understanding, that a black woman—already as relentlessly political in living out her mystical visions as Brown himself was becoming in relationship to his—should find her dreams recurrently troubled by the apparition of this white insurgent, predicting both his courage and his failure. Harriet Tubman finally was prevented from joining Brown at the Ferry only by sickness (Du Bois, 1997, 124–125).

In sum, John Brown, offers Du Bois, was a man "simple, unlettered, plain, and homely," whom "no casuistry of culture or of learning, or well-being or tradition" moved in the slightest degree: "Slavery is wrong," he said,—"kill it" (Du Bois, 1997, 173). "Destroy it—uproot it, stem, blossom, and branch; give it no quarter, exterminate it and do it now," Du Bois will apostrophize, and then conclude: "Was he wrong? No." Brown's boldness exhibited a "clear white logic"— exemplary of the kind of "prophetic resolve" before which most of us "hesitate and waver . . . now helping, now fearing to help, now believing, now doubting," and rightly so, "until we know the right" (Du Bois, 1997, 172). But then, finally, we must own our own response to the Riddle of the Sphinx ("answer or perish!"). "We are but darkened groping souls," Du Bois continues, "that know not light often because of its very blinding radiance. Only in time is truth revealed. Today at last we know: John Brown was right" (Du Bois, 1997, 172, 179). He succeeded—by walking straight forward into his own decades-long trial by wilderness and the wildness of wickedness—in putting the whole system of slavery on trial.

"Haint" Virgil

It is just such a groping after the "dark low whispering of disembodied voice"—the Riddle of the Sphinx in Du Bois' characterization—that points toward what must be engaged in our postmodern moment. Du Bois' formulation indicates a "Something" that is yet the great theological groundswell that throbs indistinctly below the feet, under the rib, up the reptilian cord of gray matter. In the nineteenth century, it gathered through "upstretched aspiration" of slave, as much "physical and mental pain from beneath" as moral calculus on high, not only cry of Abolitionist, but cry of women and black, cry of Truth, provoking thunder of Douglas, "visionary" and perceptible to the eye only as nightmarish and "damnable," the underworld making itself felt through rhythm and groan. This, Du Bois says, was the voice of Spirit. But this also, in our day, I would offer, is the terrifying whisper of fallen towers and markets, the failing schemes, the silent screams against the American Dream, percolating up through polluted ground or falling from the sky in colors unmistakably "not white." We face today, a necessary and unavoidable pedagogy of terror, and the place to turn to learn its lesson is to those who, like Dante, have in the past tracked their unwanted tryst with such, or more recently, have wrestled, like folk of color, to elaborate its tensions into vital intensities of life and spirit (see again Long's work on the *tremendum*).

I have styled John Brown a Virgil, credentialed to appear as such, for us, by his own Dantean descent in the dire hour of his time. It is important to learn from the latter's account that the clear visage of such a guide emerges only at the nadir of fear. Masciandaro marks the moment carefully in Dante's own initiation. It is not when the pilgrim first utters a *Miserere di me* that the Silence congeals into a recognizable face (Masciandaro, 15–16). The cry is important—pushed as Dante is, by the apparitions of evil incarnate (leopard, lion, and she-wolf), beyond his easy nostalgia for a nice garden ("suburb") of innocent creatures, to confront "the terror of separation from the natural order," and finally face, alone, the fact of "insurmountable evil" (Masciandaro, 11–12). These are beasts simultaneously of imagination and matter, both natural and metaphorical, looming large as the intimation of Evil Spirit itself on the fevered eye. Dante spies a ghostly outline in the heat of his fear and cries out. For Masciandaro this is enough to begin to materialize a shade of help. But this help is still uncertain, shadowy; it appears ancestral and archaic. The intuition invokes whatever "it" may be, for aid. Extremity, we might say, is not choosy. Only through willingness to solicit "Otherness"—before it has revealed "identity"—does guidance finally come (Masciandaro, 13–14). This "zero degree" of irresistible opposition, of "tragic condition" and its revelations in the dark, is the *sine qua non* of conversion for Masciandaro. His analysis is consummate.

This one who "through long silence seems faint" to Dante only begins to take on resonances of voice once the wayfarer is able to begin linking his own inner silence with the silence he projects onto the "object" that begins to appear before him. This one has for long been a *vox clamantis in deserto*, says Masciandaro, " 'silent' only because no one would go to such a barren, terrifying place—a place of death—to hear him (as Dante the wayfarer is learning to do) in this scene" (Masciandaro, 14). (We who are white perhaps begin, here, to hear a voice from a repressed, forgotten, scene of old Virginia, or indeed, from the ghetto, from the *barrio*, from below the equator.) Dante simultaneously renames the space of his encounter a "vast" (not a "little," as he had named it before) desert. (The "Third World," we need remember, is really more than two-thirds of the world.) It is this starkness of the perceived "surround"—no longer colored by nostalgia but populated rather by images and horizons of fear—that clarifies the sound. The stripping has begun. The place—at once imaginary and real—is a "place of absolute fear" (of death). It will be followed by more such perceptions—indeed, wave upon wave, circle upon circle. This is not a sudden insight for the sake of avoiding what is thereby glimpsed. Dante will have to look, and walk, straight into what he fears for quite a time. The response, necessarily says Masciandaro, is a cry.

And it is the cry that both separates and joins Dante and his deliverance. The moment is "hierophany." Virgil has not yet been revealed as such, but only begun to come forth as shade from the past, as mask of the neglected other, as forgotten ancestor—now allowed, wanted, absolutely needed (!) as comfort (Masciandaro, 16). The absoluteness of the terror provokes a turn to *anyone who can act as God*, says our commentator. (For white teenagers, staring straight through the American Dream, perhaps Tupac Shakur.) And the act is by way of contradiction—a human appearing as manifestation of divine, at once utterly the same and utterly different. In extremity, the edge of paradox does not cut, but consoles. The voice identifies itself immediately as the poet (Virgil) of a former justice-seeker (Aeneas, who himself will engage an underworld journey), fleeing a burning city. But the apparition is quickly depicted as offering a kind of "consolation of art," counseling for Dante the climb up the delectable mountain (the earthly paradise, devoid of evil and tragedy) to the sun of happiness without a trace of obstacle (Masciandaro, 18). The counsel is but another test, however, part of the rhythm of oppositional scenes that organize the dialectic of deliverance into a gathering recognition that the only way up is down. Neither nostalgia (for innocence) nor the analgesic (of poetry) can go bail for the descent.

Dante, interestingly, demurs. He salutes Virgil now so-named, as the fount of his own fair style that has been internalized through long study (Masciandaro, 20–21). The poet-apparition is placed as internal to the authority of Dante himself, a kind of spirit-possession that animates his own artistic vocation. But here Dante's artistry must be made to serve mastery not as "aesthetic" but as "sage." (What Eminem has yet to learn.) He points his own Virgil toward the incarnate evil that has driven him back into the dark wood, stricken in his own body now, with the very trembling that had earlier merely tinged the air around the animals (Masciandaro, 22). The fear has indeed, in Hegel's sense, gone "absolute," shaking the spirit to the core. No mere speech, no "fair style," in and of itself (and no theology, simply by that name, we might add), can exorcise *this* presence of evil. The poetic counsel, answering as Dante weeps, points the necessary direction: "this beast lets no one pass" (precisely!, we might say, in regard to the way whiteness tries to work its privilege by licensing avoidance: there is no "passing," even for "whites"). The remedy is an "other" journey.

The Pedagogy

But the way of that other trek has already begun. Masciandaro is at pains to point out the pattern. Virgil's counsel has emerged only at the point where the drama of opposition has gone internal to the protagonist and excavated space for such otherness. Kenneth Burke supplies Masciandaro the particularities of the requisite "walk." Burke speaks of discerning a kind of "tragic grammar"

> behind the Greek proverb's way of saying "one learns by experience": "*ta pathemata mathemata*," the suffered is the learned . . . A *pathema* (of the same root as our word, "passive") is the opposite of a *poiema* (a deed, doing, action, act; anything done; a poem). A *pathema* can refer variously to a suffering, misfortune, passive condition, state of mind. The initial requirement for a tragedy, however, is an action. Hence, by our interpretation, if the proverb were to be complete, at the risk of redundancy, it would have three terms: *poiema, pathemata, mathemata*, suggesting that the act organizes the opposition (brings to the fore whatever factors resist or modify the act), that the agent thus "suffers" this opposition, and as he learns to take the oppositional motives into account, widening his terminology accordingly, he has arrived at a higher order of understanding. However, this statement may indicate more of a temporal sequence than is usually the case. The three distinctions can be collapsed into a single "moment" so that we could proceed from one to the others in any order. (Masciandaro, 24; Burke, 39–40)

Here the pedagogy finds its ongoing completeness. Dante has been driven to the desperate quarters of his own topography (internal and external) where travail takes no captives. For multiple reasons—in political exile, in the grief-grip of never consummated love (of Beatrice), in the lost-ness of mid-life—Alighieri has been enervated by an anxiety he can in no wise exorcise. The drama he unfolds acts by its edges, piling up contraries at each turn of scene until the percussive intensity has cut all the way under the skin. Is this mere fantasy, imagined itinerary for the hallucinogenically deprived? If so, it nonetheless has interwoven politics and perdition, terror and titillation, into a trek that teaches. At the core is the role of fear and its frictions. The task is intensification—the mobilization of defenses against themselves by way of an alteration of invocation and denial, solicitation and repudiation-by-its-opposite. Burke's benchmark spells out the grammar: it is tragic by way of embraced suffering. Action invariably—unavoidably!—calls up its counter. But here is the hallmark for a whiteness on the way to withering: it must embrace and suffer the very "blackness" it has historically called into being and bound to a curse. John Brown marks the way.

John Brown can well be imagined as the shade who begins to appear when white men (especially; white women may well develop a different nuance of such counsel) of our day are brought face-to-face with our own peculiar "beasts of appetite" and run terrified into our various valleys of despair. (And here the discourse necessarily "slides" in its metaphor: it remains nearly unthinkable for most moderns that conversion in the twenty-first century might entail a reschooling in the harder ways of nature and of hunter-gatherer vulnerability and joy—although books like *Ishmael* and movements like anarcho-primitivism are seriously raising precisely that question, and *Ishmael* itself does link the question to racialization by speculating that the mark of Cain, as the prototype of expansionist-agriculturalist-exiting-garden-life-for-good, may have been "whiteness." I do not *not* mean that—that may well prove to be part of the teaching of the environmental crisis we find ourselves ever more tangled in—but that is a slightly different focus for a different argument. Suffice it here to suggest that our current social conflicts and economic predations present beastly enough visages for a beginning experience of terror.) Of course, such a ghost (as Brown) is not graced with recognition. The numbers of white men who could even place the name if given it would surely run to less than one in ten.

But the voice begins to speak anyway, without a name. It counsels return into darkness—not merely metaphorically, but historically, into the very modern advent of "darkness" itself—a strangely "white" invention by way of projection, that has brought so much tragedy and trauma in its wake. Brown counsels whites to return to "the black"— and here the meaning is multiple: a metaphor, marking a position, giving rise to a culture. Of course, I am speaking simply in order simply to speak. The metaphor historically designates a proliferation of positions, attended by proliferating cultures. Blackness is no one more one thing than whiteness is.

And yet there is a generalizable truth, a useful fiction, in the binary here. White/nonwhite is arguably *the* organizing trope of the globe—a fictive difference leveraging very real resources and repercussions over a 500-year career that has yet to be effectively disrupted (Mills, 137–138). The binary morphs continuously and in its shifting shapes hides its violent efficacy. Except from those on whom it falls. People of color know (whether from South Central or southern Afghanistan). If whites would know, they must teach themselves by way of those who do know. Thus, John Brown teaches—a white shade of a dark night who breathes in the rhyme of delirium, of war. And this also he teaches: it *is* war (even if I should choose to fight nonviolently), and never has not been such. Brown as a new Virgil counsels me as a new Dante: be prepared for the kind of struggle that is signified by the word "war": embrace the terror, the wild, and be taught by such; go to the core, the deep truth, the hard kernel and do not pretend to resolve the villainy with a word, without blood. At the least—I must face the blood on my own hands, on my face. It is not mere metaphor. The clothes I wear are scarlet under the infrared of spirit. They are not innocent of what has been shed around the world in their procurement at prices I like. The body I inhabit is burgundy-white, full of dried blood of all whose dreams have been buried with them at the far end of my privilege and the gun I employ to ensure it. I eat others. I am born of their sweat and carved fat, their crushed life chances, their forced labor and raped vaginas and starved kids and untreated diseases and early deaths. My earthly destiny of delight depends upon theirs of damnation. And there is a calculus in the making.

The Possibility

In simplest terms, in America, *beginning* to speak about race—after ceasing to speak in order to first undergo the necessary journey "down"—would mean calling for a reduction of the average lifestyle by four-fifths. As less than 5 percent of the world's population consuming more than 25 percent of the world's resources—in a world where fully one-fifths are perishing for lack of access to enough resources simply to survive and where everything, literally and concretely, is connected to everything else—four-fifths of the American take is "stolen goods." And the sense of entitlement to such (and outrage if access is merely questioned, much less interrupted) without bothering to imagine the bloodshed has everything to do with skin color. Until white theology understands that white privilege and power and plumpness is fundamentally an embodiment of the life substance of others of color around the globe, and resolves to disenfranchise whiteness for the sake of a broader and more just circulation of resources and life chances, there is really nothing to be said.

Such a radical redoing of white identity and expectations cannot even be imagined apart from a shaking of white "being" to the core—a shaking that cannot be accomplished simply by remaining in one's (white) room and "thinking thoughts." Ultimately, it can only be accomplished as a "grace from without," undertaken as a process of initiation ("baptism")—economically, politically, socially, culturally, spiritually—back into communion with the other whose economic exploitation and social rejection constituted the consolidation of white identity in the first place. That passage can only be accomplished by "passing through" the "black" experience of the other—yes, through study and meditation and self-examination, but even more importantly, through encounter in situations where those others have majority power and whiteness can thus begin to be *experienced* as the minority identity it really is globally and *owned* as the mythology of supremacy that the majority have rightly grown to hate. It is fundamentally a matter of "exorcism"—of excising the person from the presumption, from the political inscription *in the body* that has engrained social superiority and cultural normativity as an unthought birthright. The hallmark of that displacement will first of all be the experience of terror and loss—of position, of privilege, of power, of identity—as the precursor to new identity, new position, new vulnerability. This is the meaning of the "new birth" admonition given to Nicodemus that was cited in the beginning of this chapter. As should be evident from more than 500 years of history now, anything less radical just ends up being used as more ideology for more of the (white) same.

This then is the deep meaning of whiteness that remains to be understood and disavowed. It is a supremacy of entitlement. And it is a lie. It cannot be saved, only foresworn and fought against, in lifelong struggle. John Brown, in his day, dared strip the lie to the bone. He found all bones white when shorn of flesh, and all blood dark when shed. There is no escape. There is only the way one lives, and what one lives for. This is his theology—the only possible white theology that is not a curse. In John Brown, the Sphinx speaks our own peculiarly modern "vision of the damned," our "groundswell of Spirit," of Africa and Asia and aboriginal America returning from the soil like blood-of-Abel crying. Not hearing is not an excuse if I understand auguries of the last day relatively rightly. The response Brown coaches begins with the return Dante exhibits. It means allowing ourselves to be led before the terrifying apparitions of appetite that underwrite what passes for (our) "civilization," to be brought face-to-face with the dead we "consume" daily until their cry exercises our throat, their groan becomes our breath, their weeping our trembling. Any word—even a theological one—that pretends to preface such is already a lie. A white one.

Thus, I would say simply: the condition for the possibility of white theologians beginning to speak of race *cannot* be simply *said*. It implies a torturous, life-long journey that can only be adumbrated, not spelled out. It means return to the *arche* of American history, to the memory so carefully forgotten—of red skin subjected to ruse and rage, of blood spilled and gold taken, of land pillaged and village burned and children forcibly removed from family, of residential schools and reservation rules and raids of resources and culture continuously; of brown skin subjected to wholesale "blackening," of ankle and arm shackled and body sold and surplus enjoyed, of neck noosed and knuckle severed and cotton claimed for a cheap dime, of gullets garroted with cheap wine, lungs tobacco-ed, hopes sold dope, heads cracked, kids criminalized into the prison system that is now the largest employer in the land; of yellow skin and tan skin and olive skin all likewise "skinned" and scanned and scammed until the white man emerged from the dark age as the new titan of the world-clan, building towers, boasting whiskey sours, ripping flowers from their beds for the sake of the concrete castle of fiction called Wall Street. How does one go back through all of that? I can only venture that until the resources are shared, any word spoken will be more smokescreen. And until the white body writhes with red rage, until the white heart heaves with black tremors, until the white head bows before yellow dreams and tan schemes and olive screams for a different world, any communion claimed will be a contrivance of denial. A theologian—speaking of resurrection, in a body *not* bearing the scars of its own "crucifixion"? Impossible!

The answer to the question is quite simple: White theology must learn how to *stop* speaking in the key of presumption and habit and lifestyle, and begin to speak explicitly only to the degree it *is* learning to weep consistently "in" the very hell its tragic history has created. It can enter that hell only by shrinking.

8

Anti-Supremacist Solidarity and Post-White Practice

For nothing can be sole or whole
That has not been rent.

—*William Butler Yeats*

While it constitutes an overwhelmingly complex discursive event, the 1992 LA uprising could be construed as, in part, the eruption of a kind of urban commentary on official "white" speech and authorized consensus represented by Simi Valley. It was a moment of shock for the nation, when racialized reasonableness and racist lawfulness showed their illogic and the communities that have most painfully suffered that illogic in history ritually exhibited its contradictions. It begs to be read under the rubric not only of a momentary "black"-led attempt at "urban exorcism" (violent repudiation directed against some of the sites of economic exploitation and symbols of political constraint in the inner city), but also of continuing white disingenuousness. The aftermath reveals how a pervasive power formation reconstituted its hegemony after a moment of compromising self-revelation and refused a possibility of education. What could have served to challenge white self-certainty in the end only further conflated normative whiteness with business-as-usual.

To wit: the first trial related to the beating of Rodney King convened numerous audiences beyond its suburban courtroom. At stake in the "speaking" represented by the trial proceedings, and especially by its courtly pronouncement of a verdict was not only the issue in contention, but the legitimacy of the judiciary system and the authority of the policing function that constituted its point of contact with the street. Pre-April 29, 1992, the audience roughly delineated as "inner-city LA" maintained a kind of public silence, with vigilant ear cocked toward the place of authorized speech. But very quickly upon hearing the verdict, this audience retracted its tacit willingness to act *as if* it accepted the authorized discourse and actively ritualized, through the medium of fire, an intensive disavowal. While certainly shot through with all kinds of individual motives, and intergroup, interclass, and interethnic cross-currents, its overall expression could be read as the collective voicing of a counter-verdict, specified in the attacks such as that on Dennis Rodney's body, targeting a much wider audience in part convened by various media. The burning represented, among other things, a spontaneous (though temporary) reassertion of popular voice and local agency in a domain suddenly compromised in its ability to continue to stage "official" authority. In one sense, this collective outburst forced reconsideration in a second trial.

But what was reconsidered during that second trial remains ironically occulted. According to a *New York Times* account of the proceedings dated April 24, 1993, the jury based much of its decision on something only adventitiously noted, but not emphasized as significant, in the courtroom itself. The video of the beating of King showed, in one of its sequences, a baton blow to King's body while he was lying supine on his back, entirely passive—apparently a kind of gratuitous "hit" unrelated to any observable movements of resistance on his part. Having earlier dismissed the expert testimony of both sides as reciprocally tendentious, the jury finally felt able to affirm (unanimously) a violation of civil rights in large part on the evidence of this single "unwarranted" blow.

And it is possible then to read the admittedly multivalent meaning-structure of the second trial as follows. If Officer Powell is not convicted of intent to violate "rights," then Officer Koon's supervision of the arrest process is not culpable either. At issue is the "rightness" of authority *vis-à-vis* the "rights" of an individual citizen, complicated radically by assumptions, perceptions, representations, and historical memories of the meaning of race in this country. In the video evidence detailing the way the disciplinary technology of authority was "brought to bear" on King's body, a single image of a blow to a paci-fied black body becomes the image that carries the day. What appears finally to be a moment of critical clarity—that writes Rodney King legally into the American (white) discourse of rights as "victim" and valorizes his popular construction in the inner-city (black) discourse as "martyr"—is this one blow.

But ironically, the verdict thus attained only further entrenches the racism. What meaning does this verdict constructed from a single blow emblematically reinforce from the perspective of a question about the hegemony of white racial domination? Certainly not resistance! What is ultimately "licensed" here is the image of powerlessness. The verdict voices a judgment that white power, as long as it already has the racialized body lying supine on the ground, shall not also strike it. Among other things, it represents a concern not to overdetermine the public posture of authority.

In the single word "guilty" pronounced upon a single blow, hegemony is able to ramify and reconsolidate its authoritative position (and discourse and norms) in the very act of criticizing itself of (unauthorized) "excess." Excess is already signified in this scene of subordination by the black maleness of the body on the ground and it is already effectively controlled in the supine posture forced upon it (Butler, 18–19). If authority strikes here, the "black" excess subdued on the ground leaks back over onto the "white" side of the line. The blow becomes significant of a loss of propriety, a surrender to the "savage" underside of white legitimacy.

What is then punished, in this verdict, is perhaps not the exercise of *too much power*, but precisely the *loss of power* through a compromise of its appearance: a sliding outside the proprietary norms of hegemonic power which themselves constitute that power's authoritative style and potency. From this point of view, what is judged in the second King verdict is not the injustice of white racial dominance, but the slippage of the controlled style of a disciplinary formation, a surrender of the image of civility in a mimetic self-revelation of violence. The black male body has already been publicly constructed as violent in the metonymic displacement of historic white American (racial and geopolitical) violence; it has already been projected as a "significance of savagery" whose beginning point in white imagination is forgotten. With sickening regularity in the history of this country, white fear has simply mimed its own projections, beating its image of blackness into multicolored brown bodies. But in this case, to hit King's already pacified body one more time is suddenly to puncture the anonymity, to slide into the mimetic circuitry, and become vulnerable to recognition.

Before an eye asking questions about racism and hegemony, the King verdict itself can be understood as "really" about the self-disciplining of the hegemonic white body in its own moment of slipping out from behind the authoritative mask of civility into a recognizable form of its power to produce violence. From this perspective, it marks not so much a moment of justice as of hegemonic power ironically reconsolidating the authority of its disciplinary norms. Following the brief moments of media-certified elation in South Central celebrating this second verdict, the old veneer of silence, punctuated by sirens and gunfire, once again descended to constrain the anguish and anger that constitute contemporary urban authority's unremitting subtext. The future of that silence and its anger should not be underestimated. The irony is not lost even if the pedagogy has been refused.

Anti-Supremacist Solidarity

Chapter 8 offered testimony to a white man committing race-suicide, looking his own eye in the mirror without flinching, acting over the course of two decades to find the vulnerable place of white violence, where pressure could force disclosure of the real character of white power and precipitate change. The quest was not merely a concern for justice for suffering others, but finally a risk of the whole person—family, heritage, name, future, wherewithal, limb, and loin—for the sake of an ultimate disposition. John Brown was wagering on the meaning of salvation in the midst of thoroughgoing corruption and all-encompassing violation. Whether the choice for counterviolence was the only possible choice in such a scenario of desperation remains debatable; that resistance required risk of everything and utter intransigence in refusing complicity is not. The body had to be put on the line, in the way of the bullet, the bombast, the entire behemoth of chattel slavery. The "no" to be spoken required backing with blood, if necessary. Nonviolent resistance on the order of a Martin King or a Mahatma Gandhi would have suffered its own to be shed; John Brown, though he did opt finally to kill, knew also that his was already puddled on the ground even as he acted.

The word from the wild underside of our history is that race is code for war, a war that has not ceased even if its front lines have shifted from field to factory, hollow to housing estate, mountain height to suburban mall (remember the black man who perished under security guard knee in Detroit's Fairlane Center in 1998?) and its low intensity everyday-ness now erupts into blatant conflict (mortar fire and bomb-drops and missile strikes) only at a great distance from its current domain of privilege (except when dark-skinned others dare reciprocate the terror against a home-front tower). One film reviewer of the recent release, *Cold Mountain*, tracing poor white experience of the war John Brown precipitated, wrote rightly that the Civil War never really ended. It merely morphed into multiple theaters and has continued in other struggles. The question for this chapter then is "how to fight?" What can a white person today do to be true to the history, enter into solidarity with its sufferers, and help forge a different future story? And what is at stake in the struggle? I do not have simple answers but only provisional parameters for practice.

Dilemmas of Solidarity

The first concern has already been addressed. The deep question is how to gather enough passion to battle for a lifetime. There is no quick fix, no easy remedy for a conflict that has organized a half-millennium and encompassed a globe. The first task is not to flinch, not to knee-jerk a demand for something easily do-able, to treat the dilemma like an accounting problem. It *is* an accounting problem at one level—a debt of between 1 and 5 trillion for unpaid slave-labor alone, much less the perhaps 2 trillion more for underpaid Jim Crow wage work and industrial extractions,[1] or an as-yet uncalculated "take" from the present day prison-industrial complex whose for-profit stock returns and use of youth of color as a cheap, captive labor force are becoming notorious (Schlosser, 51–77). All the wealth so generated over time accumulates largely where the affluence is gated and circulates primarily where privilege is taken for granted. In such a view, whiteness in general and white maleness in particular are living forms of social indebtedness—of such immensity, most of us simply shrug our shoulders and refuse to even try to imagine what redress could possibly look like. The long-standing and constantly renewed movement for reparations for slavery—whatever its knottiness as a practical remedy—does mark an important piece of the meaning of white repentance today economically and psychologically. But if we who are white bother to question the source of much of the substance we enjoy, we quickly find ourselves in the grip of guilt and eager to pronounce our own absolution ("well, certainly my family did not own any slaves"! "My parents came over from Germany and worked hard for all they own!"). The need, however, is to stay open to the emotion and learn its lesson.

The Pathos of the Predicament

The first task is not to flee the feeling. Race is a "problem" for which there will be no quick fix solution. What is demanded rather is integrity in owning a long history and a deep perfidy. Quick fixes are themselves indications of a positioning in power, a habit of applying intention to irritation and mobilizing resources to eliminate the source of pain. This pain, unfortunately, is not going to disappear overnight or without sustained and costly challenge. The witness from the side of the suffering is that the infestation is integral and will take utter immersion in "the problem" to give birth to solutions. It is part of the meaning of privilege to relate to the world as "plastic" and malleable to our desires to alter its appearance or solve its problems, as we have seen in chapter 6. For most of the world's populations—and certainly for most blacks in this country for most of its history— modern reality has been unyielding, unforgiving, hard, and invasive. Recovering some sense of common interest and identity with the rest of the planet, some real inkling of solidarity in suffering and joy in surviving means refusing the impulses of denial and avoidance.

The problem of race is as deep as the body we "are," as subtle as the culture we inhabit, as long-standing as the modernity we take for granted as our mode of being. Resolution of the problem will take more than a lifetime, perhaps as long a time as the problem has been in the making. Most whites engaged in the Civil Rights effort of last century mistook the malaise: after offering a year or two or even twenty, they cut their hair, wiped their hands, bought their split-level, secured their portfolio, and considered any remaining difficulty a fantasy or pathology of the ingrate "other." But the infection—like any bug treated with only half a course of penicillin—merely mutated its resilience and dug deeper into the (national) tissue. In the metaphoric of illness, this disease is epidemic, the threat terminal, and the treatment regime lifelong.[2] What matters is a sober prognosis of the sickness in the body politic, a decision to battle long term, and clarity about the kind of resistance one is most equipped by personal history and cultivated ability to offer. Small victories may well be anticipated and enjoyed, but large-scale change will be (as it has been historically) slow and incremental. The demand of integrity at this point is simply to "go down fighting." As Francis Moore Lapeé, author of *Diet for a Small Planet* and crusader against world hunger, once said, "If you are struggling over an issue that can be solved in your lifetime, you are probably wasting your life."

The Conundrum of the Condition

The spiritual meaning of such a rite of passage as imagined in chapter 7 is that of exorcism and idol-breaking. The conundrum for those of us who are white is that the idol is under the skin and the demon is in the mirror. This chapter outlines the consequence of such a self-confrontation. Theologically it can be summarized as an issue of soteriological combat: the need to repudiate a false figure of wholeness and embrace a form of integral "disintegration" (Esau's brother, we might remember, had to exit his family name and wherewithal in order to rediscover himself as Jacob-Israel in a process of painful self-recovery after years of self-exile). The conundrum of race is such that there is no salvation for whites as white and there is no solidarity with others except as white. Neither exemption from the history nor captivity to the identity can be tolerated. Learning to confess guilt in responsibility and simultaneously embrace the teachings of uncertainty (as the only way to become more than just white) are both requisite. The demand is double and damnable.

Inexperienced white students of mine in the seminary where I teach have sometimes told me, after attempts to engage with African American students, that they feel "damned if they do and damned if they don't." If they try to initiate conversation with questions typical of white protocols for greeting (e.g., "Hi! What do you do for a living? Where do you come from?"), their questions are often met with a cold shoulder or a vague answer (as questions that black culture often identifies as highly personal and inappropriately invasive for strangers to ask). Invocations of "color blindness" in classroom discussion ("I don't notice race; I just see human beings") will likely arouse an undercurrent of suspicion, since the behavior and lifestyle of the whites offering such comments almost never matches the claim. On the other hand, attempts by whites to learn about or even participate in black culture (e.g., adopting street lingo or gestures of greeting, learning how to rap, etc.) may result in rolled eyes and exasperated mirth, or more seriously, in accusations of imitation and disingenuousness (being a "wigger" or even stealing the "home product")—though not participating only reinforces the sense of difference and perpetuates white ignorance and aloofness. Such initiatives can quickly "deadend" in a sense of frustration, of feeling caught in a double bind. And I respond: Precisely! The paradox *is* the pedagogue.

Living under the regime of racialization and being marked as "black" (or "brown," etc.) has always been exactly the dilemma of a double bind. White supremacy in practice has ever meant an impossible demand for nonwhites around the globe: being held accountable to a standard of white achievement, Western development, university "enlightenment," and elite refinement as the measuring stick of worth and access *and at the same time* being refused full (or even any!)

participation in the white-controlled institutions and corporations that inculcate and offer such skills and possibilities precisely because the skin still shouts "otherness." Until no one has to suffer such a double bind anywhere, everyone should labor with its intricacies and ironies everywhere. Such is simple justice as well as the only path to mutual understanding. Any desire to be exempt from the tiny bit of tension whites might experience in such interactions (not even in the same galaxy as the levels of violence and the costs of exclusion blacks regularly suffer!) is the very meaning of supremacy. For whites, the pain (however slight) of such an experience is one of the gateways into learning about the injustice.

Stated more programmatically, for white people, the passage into a genuinely multicultural democracy necessarily *passes through* a form of "racial conscientization." There is no remedy for ignorance except learning, and no learning without facing one's ignorance. Those of us who are white can become genuinely pluralist in our ways of handling ethnic encounter only to the degree we become racially self-conscious. To embrace what already is the case in this country, we will have to become far more comprehending about what has been. In effect, I am arguing that white antiracism requires an approximate re-racialization at the same time that race itself is made the object of a new abolitionism. The ideological emptiness of whiteness must be filled in with its historical *pathos* and existential ignorance and present insouciance even as a more theological and ethical self-emptying is engaged.

The Problem of the Pedagogy

But the deep question that this double demand (of anti-supremacist solidarity) raises in an urban seminary (such as the one I teach in) is not first of all one for white students, but for white faculty. The basic argument being slowly assembled here is that a "seminary"—as indeed any other Western academic institution—by definition embodies a historically European cultural *ethos,* not only in what it teaches as content, but in the pedagogy it practices as habit. The colonial Christian presumption of access to a superior level of ultimate "truth," compared to other religious and cultural traditions, did not simply stay put at the level of perception and cognition. It also encoded itself in deep-memory structures of posture and motor-skill, in modes of meaning-making and regimes of reality-testing that are quintessentially ritualistic. A seminary, whatever else it may be institutionally, gives structural articulation to the prerogatives of a particular way of prosecuting truth. It embodies an epistemological presumption privileging academic *argumentation* and a pedagogical practice based in disciplinary *compartmentalization* that is both socially specific and culturally limited.

When students from the urban streets of millennial America step into the classroom of such an institution, they are confronted with an implicit demand to become bicultural (or more accurately, "tri-cultural," since merely surviving in America already requires a certain level of bicultural competence). It is not only a question of mastering a new content, but of developing competence in a new calisthenics. Academic process is ritual "congress": the skill is not simply a matter of mobilizing information, but of "liturgy." What qualifies as truth has as much to do with style as with content, with dispositions of the body-in-motion as with texts-on-a-table. "Truth" is not "one" in this argument, but is rather a function of a set of practical presuppositions and policing powers that always invoke a specific cultural code, as Foucault has argued (Foucault, 1984, 51–75). Students coming out of cultural formations that privilege the oral, the kinesthetic, the perfor-mative, and the personal must come to grips with a whole new panoply of ritualistic regulations and rhythmic restrictions. This implicit institutional demand for *re*-cognition (by students) at the deep-structural level—of what *qualifies* as "knowledge" in the first place, and how such knowledge is *socially policed* and *normatively adjudicated*—functions to encode supremacy if it is not reciprocal. If faculties presiding over such a cultural theatrics are not holding them-selves accountable to becoming biculturally competent "in kind" in the performative traditions of those they are presuming to "teach," whatever else they may be teaching, they are in fact, teaching a version of cultural superiority. The very structure of the relationship in such a classroom encodes a one-way demand that cannot *but* communicate, in this society, a meaning of white supremacy. It is probably impossible in terms of time and energy to internalize deeply more than one or two other cultural languages, but the experience of one such attempt at "crossover" can be enough to begin opening up consciousness of "the supremacist body."

Such at least is my thesis here. A seminary, concerned with notions of truth at the level of things "ultimate," animated by a belief that such an ultimacy was and is actually found in human flesh, *must* attend to its *own* flesh. The very idea that the primary focus should be on a truth that is "believed in the heart and confessed with the lips" (Rom. 10: 9) is itself the subject of a very particular project of *embodiment.* In history it is the project we euphemistically call "modernity": the global intention of aggression that has, as its dominating set of interests, a body inhabiting a cultural "safe house" that continues to be policed as "white," "male," "heterosexual," and hungry to swallow the whole world. In our day that consumption continues apace. The fact that other kinds of bodies are now climbing up inside that house does not comfort. The question of whether education can be done in service of arousing a different kind of intentionality is almost hopelessly utopian in its very articulation. But the insistence of growing numbers of non-European peoples for something like "hermeneutical parity"[3] in many of our educational institutions in recent decades does raise the right form of deep question. "Reciprocity" is what wants bones to live in, inside of our pedagogies. The demand is for a skill that is actually prototypically "African": facility with a kind of "syncopation" in learning to express one set of rhythmic structures while simultaneously keeping track of another (or indeed, interweaving between the two when appropriate). The demand for myself and my white faculty colleagues is to become socially and culturally and habitually what we already are economically and subconsciously and insidiously: inhabitants of more than one culture.

Politics of Solidarity[4]

Given some of the more subtle issues a quest for solidarity raises, what are some of the more structural concerns such an intention must confront? In taking public responsibility for its historical position, a self-consciously "white" search for greater wholeness must be both self-referential and other-oriented. In calling attention to the constant rearrangement of resources and power in favor of the "haves" of our society, white self-confession requires more than mere self-naming or "me too-ism." It demands clear steps of conversion away from the historical intentions and material privileges of white self-interest. Such a conversion requires a kind of hermeneutics of contraction, able to analyze privilege in all of its historically specific formations. White supremacy is not the same in working-class articulations as in middle-class institutions. Nor is it the same over time. It operates differentially across gender lines, within religious traditions, in-between sexual orientations, over the course of generations. It cannot be simply comprehended and countered as a monolithic social formation. And it is shifting its shape rapidly and radically on the contemporary political landscape as we begin the new millennium (Klor de Alva, Shorris, West, 59). The new dominant minority on the home front is now Latino-looking and Spanish-speaking, and the new enemy of choice geopolitically, Middle Eastern–appearing and (presumptively) Islam-believing—with all the complicating political ramifications such shifts will necessarily entail. Not least will be a whole new set of exploitable fault lines between disadvantaged groups that will undoubtedly give rise to new "divide and conquer" strategies at the top levels of society.

The Parameters of Place

As already cited earlier in this writing, it seems likely that the great divide of the future will become increasingly spatialized—enclosed communities of affluence defended by high-tech security systems whose profile of "membership" is not exclusively decided by race, but rather by a combination of acquired characteristics defined overall by technocratic norms of competence in symbol manipulation and access to data banks (Rifkin, xvii, 176). Dark skin-color, alone, will not necessarily exclude[5] as long as the other signifiers of middle-class belonging (clothing styles, speech patterns, modes of transportation, networks of professional association, etc.) are in place.

In such a future, "blackness" will likely continue to take its primary range of significance (in dominant culture discourse) from the conditions of living in impoverished urban spaces outside of (and largely invisible to) gated communities of affluence. Pressure will continue to fall on the black middle class to try to secure its (uncertain) place in this arrangement by demonstrating its distance from these "thicker" meanings melanin is made to carry (such as "unemployable," "uneducated," "promiscuous," "teenage mother," "drug dealer," "gang-banger," "purse snatcher," etc.). At the same time, privatization of the prison industry will likely continue (subtly) to reinforce the need to find these meanings "verified"[6] in actual dark-skinned people. In many communities now, the availability of both jobs and cheap labor depends on it. This version of the future also requires white-skinned poor people (as indeed, lighter-skinned ethnics of various backgrounds) as buffers between criminalized (ghetto) blackness and enclosed (affluent) whiteness. Their presence serves as evidence that poverty can be survived without turning to crime, reinforcing the idea that dark skin-color is the cause of criminal behavior and legitimating the arrest of black and Latino youth largely on the basis of appearance.

While nothing in the picture just sketched is presupposed as a form of conscious conspiracy or inevitability, it does take shape as the likely social valence of racial appearance in the new millennium. One obvious and quite thorny question for conscience—especially that of white people, but really for everyone—is the morality of living in spaces that enshrine in their very architecture of luxury and infrastructure of sprawl, surplus labor values coercively extracted from others elsewhere. For some, refusal to live "middle class" in lifestyle and "gated" in place is a deep choice of spiritual formation and a first step toward social solidarity with the marginalized (though it remains fraught, as the very ability to make such a choice in the first place is an index of privilege among those who have no choice about where or how to live). But in any case, white people need to recognize where and how wealth concentrates and how skin is made to articulate with the system of signs by which legitimacy of that accumulation is fashioned into public consensus and decide a posture of committed response towards the injustice. The counter-word to this conflation of skin color and fortune (or lack thereof) is undoubtedly "struggle."

The Priority of Struggle

Historically white (blue-collar) workers as well as white women have shared a whole set of economic interests with the black community in general. Labor unions have (at times) been able to struggle for concerns common to workers otherwise polarized by racial identification; affirmative action policies have, in fact, benefited white women even more than blacks. The search to uncover histories of struggle against domination and exploitation of all kinds, to clarify the particularities of the struggles involved and thus of the structures and interests struggled against, and to link the various histories of those efforts in a broad-based cultural ethos privileging such struggle as the paradigmatic humanizing quality across all the registers of human difference, is the task of conscience in the twenty-first century. The key here is "particularity."

Struggle is inevitably very specific to the particularities of context and thus to the problematics of the kinds of oppression historically characterizing any given context (theologically, one could say that struggle is profoundly "incarnational"). What remains vulnerable to the "divide and conquer" manipulation by the powers that benefit from oppression *could* become the source of coalition and mutual respect (as briefly happened in late nineteenth-century Populist Movement cooperation between poor blacks and poor whites before it was dismembered by planter-class manipulations of racial fear). Efforts to transform suffering into viable forms of resistance, to trans-figure constraint and stereotype into means of survival and codes of creativity, demand appreciation wherever encountered. The great temptation—usually encouraged and exacerbated by the structures of domination—is to try to evaluate such suffering, and the struggles it gives rise to, in a comparative frame. Disadvantaged groups are invited to dissipate energies and confuse priorities in competing with each other for the "privilege" of claiming the greatest victimhood. The need is rather for a contemplative eye that is quick to see pain and pas-sionate in appreciating resistance wherever encountered. The recogni-tion of "other" forms of human dignity hammered out in other historical crucibles of struggle can itself give rise to a renewed resolve to identify more clearly and engage more concretely one's own partic-ular battles.

For instance, one of the personal effects of my own investigation of African American resistance and creativity has been a sharpening of my appreciation for the Irish side of my own ancestry and a growing affinity with certain forms of Irish endurance and insurgence even in the absence of certain knowledge of how my own Kelly ancestors actu-ally lived (historical records, of course, are the prerogative of history's victors, not its victims). I now read of the struggles against British domination during the potato famine of the nineteenth century with a certain sense of personal pride. I also listen differently now to Polish friends who can spell out for me the precise way "Polish" humor was used with corrosive political effect or Hungarian students of mine who recount the way poetry kept hope alive in desperate circumstances.

There are complex and compelling histories of struggle resident in the labor movement and the woman's movement, the Civil Rights movement and the voting rights movement, in Native American efforts to resist massacre, Chinese American efforts to survive labor gangs, Appalachian efforts to survive mines, Japanese American efforts to overcome the memories of the camps of World War II, Filipino and Chicano efforts to organize against exploitation in California, Jewish American efforts to anathematize "holocaust" anywhere,

Palestinian American efforts to end the exploitation of relatives in Israel, Chaldean American efforts to avoid being thought of as little Saddam Husseins, efforts all over the globe to resist being taken over by American interests, and, yes, Puritan efforts to elude Anglican repression and early colonial efforts to resist King George III. Struggle in modernity is virtually ubiquitous. Learning to honor its ubiquity, however, means precisely an ongoing struggle to recover the particularities and the histories that alone give any struggle its power.

In America, such an effort requires sustained attention to the *very particular way white interest and identity so often infiltrates and takes over* other forms of ethnic struggle. Whiteness is the invisible tag on the American Dream: unlike "Irishness" or "Polishness" or "Jewisness," etc., it has not come into being as the name of a creative form of historical struggle, but as a claim to plunder. If any given "ethnic" struggle settles merely for greater access to or a larger share of the American pie, then "success" necessarily also means complicity with white planetary pillage.

Even if, for instance, my own (Irish-German–Dutch-English) family inheritance is not immediately indebted to the asset formation arising from slave-labor, some of the economic infrastructure and developed public resourcefulness that I have immediate and relatively unfettered use of does.[7] My own enfranchisement also has intimate relationship to the ongoing appropriation of values created by the labor of a working class whose vulnerability to exploitation has depended, in part, on the threat (or actuality) of replacement by black and brown migrant labor from the south (of the United States and of the border). And it continues to be consolidated at the expense of people of color locally and abroad whose wherewithal is constantly expropriated to subsidize the priority of white well-being (through IMF and WTO policies and interventions, transnational corporation practices and prices, higher insurance rates in city centers, lower pay for the same work, banks that accept savings deposits while refusing loans, etc.). In short, the social functioning of my light-colored skin privileges my existence with various recognized and unrecognized benefits disproportionately conferred at the expense of less "congenially" endowed peoples and it thus enjoins upon me a very particular and peculiar ethical and political struggle.

The Politics of Embodiment

Consistent with the recognition of how space is constrained by race in America and how struggle is often muddled by the allure of white entitlement, habits of embodiment, as we have seen, likewise require attention as a racial battleground. Not only is the suburb not innocent as a social location. Not only is the object of political organization often simply a bigger piece of an American pie that is already an unjust slice of global wherewithal. But the body itself is the site of an ethical "investment," quite apart from its own intentionality. Long before we arrive at self-consciousness, our bodies have already been marked by our place in the system—our habits of consumption, our ways of moving through public space, our expectations of eating and speaking and dressing, our sense of safety and nurturance or threat and violence. These things do not just indifferently touch us, day in and day out; they leave a definitive signature in the flesh. They are gendered experiences of the meaning of being a body in a patriarchal and heterosexist social order (men mostly fear being laughed at by women; women, however, have to fear rape and beating, or even being killed, by men. And of course, gay men fear all of it). They are also racialized experiences, as we have seen in chapter 6.

The different spaces and communities of our varied upbringings produce different modalities of "somatic entrainment" (Mills, 44). White male middle-class heterosexual bodies—more than any other— are invested with certain kinds of power simply as a birth right and a presupposition; no work is required, no conscientizing necessary. In a world where resources and powers are irreducibly relational—where my weekly gourmet coffee is linked to peasant labor on the hillsides of Nicaragua or Guatemala, where my choice to buy designer and flaunt accessories is not entirely innocent of teenage drive-bys in the inner city over gym shoes and gold chains, where my ready occupation of the "space of speech" in most theaters of interaction is not unrelated to a socialization process that socializes many white women to defer or resist putting their bodies on display one more time for male gaze— in such a world, the body marks an ethical placement and means an ethical predisposition. The white male middle-class heterosexual body—simply in the products it consumes, the places its assumes, the postures it incarnates, the gestures it assimilates, the powers it learns, the structures it confirms—is already a "moral entity" even before it acts. As the central beneficiary of a global system of violent appropriation and concentration, it incarnates bloodletting. Simply by being itself, it communicates to its others domination and denigration. Can it be made to signify differently?

The question here is not primarily about an individual, but rather about a social artifact, a cultural archetype, a living hieroglyph, that yet has real historical effects. To the degree macrostructural social forces are encoded in the microstructural habits of the body, this normative white male body on the American scene must undergo a certain amount of "excavation" and deconstruction simply to recover the possibility of its morality. Education at this level requires dislocation—physical embrace of the contradictions of race and class and gender privilege (and the rancor to which it gives rise) in places peripheral to the centers of institutional power where other bodies have worked out other postures and potencies not beholden to the white male norm. Here we encounter the novel idea that thought is always a function of physicality. What Michel Foucault explored to a fault at the end of his life (indeed, apparently ending his life), others such as Du Bois and Fanon, hooks and Long, and feminists of all backgrounds have known from much earlier on and without choice: critical thought responds to a need to close a gap opened by violence or pain in one's experience of embodiment. Creative vision, radical decision, motivated struggle, ethical precision are rooted in the sensation of contradiction and the drive to circumscribe its inchoateness in an expression of flesh incarnate. The tongue alone is hardly enough. Indeed, the individual body, as we have already indicated does not suffice. Complex rhythm—in concert with other bodies similarly working anguish into eloquence—is the longest-standing human antidote for such inchoate eruptions. Thinking different thoughts about reality requires becoming a different body. White male immobility inside the rigidities of an unconscious supremacy cannot be undone by mere intention alone. It will take risk and vulnerability before other protocols of movement in other spaces of power if white males are to become conscious of the terror and history their bodies encode for those others. Saying such is not an admonition for either masochism or asceticism, but just a call for mutuality in engaging the pain of our common history.

White (Anti-)Theology and Post-White Practice

These questions about space, struggle, and embodiment are also a call to Christianity to take seriously its own doctrine of incarnation. In theological terms the lifelong vocation to exit whiteness and combat supremacy can be elaborated as a threefold task of exorcism, initiation, and apostasy. The *exorcistic* moment is offered repeatedly in everyday white encounters with black folks, entertained not as something the latter do "for" the former, but as a self-initiated and self-perpetuated white discipline of embracing black difference, and especially of "receiving" black suspicion, anger, and even ostracism as expressive of a necessary challenge of the Spirit, forcing deep work at the level of psychological formation and cultural habituation. The *initiatory* moment is offered by black cultural creativity—itself already an underground current inside of white identity simply by virtue of living in America, but requiring conscious exploration as invitation to the possibility of a different kind of empowerment, modeled in black reinventions of the false-self imposed upon them by white supremacist policies and politics. The summons to "own" continuously and repudiate actively the history of privilege and power exercised by white prerogative and embodied under white skin (the discipline of exorcism), and the invitation to descend resolutely and "religiously" into the opaque depths of desire underneath the ready-made categories of the culture for the sake of creating a new self (the discipline of initiation), find their double demand summarized as a lifelong practice of *apostasy* from the tribal god of white supremacy: the politics of race suicide and solidarity with the struggles of people of color for greater justice in this world.

The Possibility of a Post-White Vocation: Apostasy

Ultimately, the stakes are those of becoming *apostate*—like the early Christians *vis-à-vis* the Roman pantheon—with respect to the conforming powers that be. The real *sacramentum*, the oath of loyalty for whites, is to become a *militis Christi* ("messianic insurgent") for life, struggling, spiritually and materially, against the tribal god of white supremacy. As such, the task is ongoing, relentless, guerrilla apostasy from that god and the social order it governs. Pushed to the limit, of course, such resistance can mean incarceration or even martyrdom. (The apostle Paul, after all, ultimately paid for his championing of the rights of Gentiles to full inclusion in the economy of salvation with his life.)

Friends of mine who run the Open Door Community in Atlanta, for instance, live with some thirty formerly homeless persons, agitate against repressive legislation (e.g., laws making peeing in public a citable offense) or increases in drug co-payments implemented by a city hospital (receiving public monies) that would literally translate into the deaths of more than a thousand indigent patients, help run a facility to house family members visiting death-row inmates, and otherwise challenge downtown institutions (including doing civil disobedience in the mayor's office) about ongoing policy changes that ever-more ruthlessly exclude poor people from public life. Their actions are not exclusively determined by antiracist sentiments. But by far the majority of the people they advocate for are "of color." They are so far still breathing, but they do at times end up in jail for their efforts.

Practically, for most white people today, however antiracist conversion implies at least some measure of real material contraction expressed as a form of social expansion. It implies pursuing a more equal circulation of assets, opportunities, and power that will simultaneously be experienced as a form of real loss. Sharing control is also giving up control, at least in the moment of fear. Seen in this way, whiteness emerges theologically as a task of mourning. It involves learning to see where, in one's own life-world, whiteness as a naive (or not so naive) practice of terror intersects with one's own personal struggles with fear, and getting help in not contributing that fear toward the practices of exclusion that already structure our common social field. In a finite world, expanded forms of solidarity and political community cannot be birthed without real loss in relationship to the status quo and serious negotiation of that loss. Here, the relevant strategies are limited only by the imagination of the actor.

For instance, one could choose one's bank on the basis of its lending practices in minority communities. One could choose where to reside based on a neighborhood's mix of population by ethnicity and class. One could select a school for one's children based on its commitment to educate in ways that do not reinforce racialized divisions. There is much work to be done—both within and outside of the workplace—for forms of affirmative action that counter the *de facto* operation of policies of white male affirmative action that has been "business as usual" in most institutions in this country for most of its history. There is need to lobby for forms of political representation (even if they are, from one point of view, "gerrymandered") that do not simply recapitulate the power of exclusionary social formations (e.g., enclosed communities that are largely, though "unofficially," white). Collaborative ventures like the Industrial Areas Foundation offer opportunities for long-term involvement in community organizing initiatives across ethnic and class divides that mobilize people of varied interests in a participative process around issues of common concern identified by that constituency. There are organizations (already existing and waiting to be created) that seek to combat the conspicuous racialization of waste management practices that legally or illegally target unorganized minority communities or even Third World countries as sites for dumping (as happened recently in Haiti with a boatload of municipal waste from Philadelphia).

In focusing on imaginable changes in the circulation of assets, opportunities, and power, however, I do not want simply to sidestep the much bigger and much less tractable issue that haunts us all: the critical need for a radically different form of socioeconomic production, itself. Whether capitalism, in and of itself, will generate enough of a contradiction of its own bases of organization and reproduction to offer real hope of something less monstrously competitive (as Marx imagined) remains to be seen. Religious faith dare not cease to imagine and work with such a hope. But it must also remain realistic enough to continue to work for change on a more "human" scale of small, incremental innovation and brief realizations of alternative visions. The models here include various "cooperative" and "community land trust" experiments, CSA (Community Sustained Agricultural) efforts, and alternative credit organizations, that seek to embody different principles of ownership than either the current private or public models offer, while simultaneously granting their supporters a viable (even if not market-rate) return on investment.

Once again though, the black community bears a witness that is crucial here. Historically, white mainstream American culture has not integrated into either its vision or its practice the reality of the socio-historical tragedies upon which it is in part founded and by which it is constituted. *Tragic vision* and *defiant rhythm*, however, are gifts that African American activism in particular knows much about, as Cornel West, among others, has asserted time and again (West, 1989a, 226–235). At this level, white theology faces a task of great tension— of learning to project persuasive ideals of utopian hope that simultaneously license practical acts of compassionate realism and even "faithful" failure. In a sense, the image here is that of a white version of Du Bois's double-consciousness: a lived tension between what is and what could be, that is carried with both indignation and forgiveness and crafted into a personal politics of resilient vitality. There is a profound need for religious communities in this country to take responsibility for the specific tragedies of our common history in ways that incorporate both tragic vision and practical wisdom. There is need to dream big, act relevant, celebrate small, rage with a purpose, and weep without regret.

On a more personal level, a white theological practice of antiracist apostasy demands a new interpersonal politics. When invited to join the ranks of "the white and the right"—the legendary legions of the rational, the ordered, the civilized, and the normal—how do we respond? A friend tells an off-color joke, a relative uses the n____ word, a coworker grumbles about gangbangers and welfare queens, a neighbor laughs about walking into the "wrong" bar and, in the dim lighting, not being able initially to see if anyone was there. A little spider web of white inclusion offers itself, a subtle "feeler" for racist camaraderie. Do we play? Or pay?

One writer speaks about cutting through that little toxic cloud by claiming kinship in whatever group is being put down (Leki, 15–16). "Did you know my grandmother was African American?" "My uncle from Tunisia just arrived yesterday." "My sister is on welfare now—has been for five months." The responses are reportedly bathetic: sad and funny all at once. Coughs, splutters, apologies, corrections in profusion. But the tactic is vastly more effective than argument in raising instant issue with the fiction. There is no absolute line between white and black. At one level, we *are* all related. Go far enough back in any of our ancestral data banks and you get to Africa and a black Eve. All of us who are tempted to think of ourselves as white are in fact merely "passing"; antiracist forms of whiteness would then emerge as the practices of "recovering passers." Whiteness is a choice, at some level, to deny a mixed heritage that includes some measure of black blood, however dilute. One way or another, white people are simply "post-black" people, with all that means for dissimulation and dis-ownership. Saying such does not presume to license facile claims for inclusion in ethnic exceptionalism, or to buttress liberal notions of color blindness. It is rather to assert that the real problem of the color line, the real aberration needing explanation and elimination, is whiteness, not blackness.

It finally is a matter of choice. Whiteness has never been a stable category or a uniform community. Demographically, in this country at least, it may ultimately be doomed to disappear. But if so, it will not likely "go gently into that good night." It will rather escalate the stakes and increase the violence. Timothy McVeigh is one expression of the reaction; the bombing in Oklahoma City one consequence (Perkinson, 2000, 350). At some level, "white power" is the responsibility of all of us—white, yellow, tan, red, brown, and black—who are caught in its maw. The grading system of color that whiteness operates—standing at its furthest pole like Aristotle's Final Cause and attracting all to its privileged position without itself moving an inch—is ubiquitous. It whispers its privilege of "I'm lighter than you-ness" at every point along the scale, even inside the black community. It is one of the modern-day sites of idolatrous seduction, a place where many of us are summoned by Caesar—or Caesar's investment banker—to pinch incense to the god. Whether we choose to be included in the empire or not—to honor the sovereign and embrace the discipline, or turn to a different power—is a theological decision. But the possibility of the choice itself rests on two ongoing disciplines of practice.

The Possibility of a Post-White Vocation: Exorcism

One of the preconditions for such a lifelong vocation to struggle is a necessary *conversatio*—a conversion of life that begins with a spiritual breakout from white supremacy, experienced as a form of *exorcism* that carves down below the brain and consciousness of the problem to its *sedimentation* in the psyche and its *surveillance* in the body. For white males especially, this means undergoing a profound "breakup" of the way power and perception, entitlement and presentiment, are entirely entangled with each other, and (re) discovering other forms of embodiment to identify with. "Exorcism" implies an intervention that aims at prying the basic structure of subjectivity loose from its presupposed center to become itself an object of choice—a conscious possibility of either "identification" or "dis-identification." It is a matter of recovering a body from a habit, not just purifying a person from an influence.[8]

Once we as whites have had the kind of "significant emotional event" indicative of having actually encountered "color" at a level deeper than just our eyeballs (often through experiencing ourselves as either the object of black anger or the subject of black humor), two responses generally present themselves. Either we can reconsolidate the safety (for ourselves) and exclusivity (for others) of white privilege, power, and position, securing our body in the equivalent of a secured position (hierarchical authority) and space (suburb, a school, a workplace, etc.) whose meanings, sanctions, methods of ingress and egress, and modes of interaction we control. Or we can embrace the abrasion as an invitation to self-knowledge and allow ourselves to be initiated into the deeper meaning of our own history, as we shall see below. In this latter possibility, we would begin to confront the degree to which white supremacy is a form of cultural habituation naturalized as a patriarchal body. Until we (white males and females both, though differently) understand "in our pelvises" the way the threat of blackness "colors" our erotic fantasies and constrains our romantic insecurities, we will not yet have gotten to the real roots of racism.

Once confronted at a level that begins to be exorcistic, however, we do not thereby cease to be vulnerable to the influences of the "self" we had been before. Racism is not simply an individual affliction, but a pervasive environment and an invasive ecology that continually insinuates itself in spite of personal intention. But such a confrontation does enable us to begin to intuit the depth at which we have been conformed, and to entertain the combination of "agony and ecstasy" required to expand the "field of our embodiment" beyond simply "being white." Confronting white supremacy in oneself, sooner or later, means exploring the possibility of learning to speak in more than just one voice, to express more than just one rhythm, to reiterate more than just one culture. Such an exercise necessarily also means expanding the social field in which such a body moves, of course.

In *Black and White Styles in Conflict*, Thomas Kochman relates the experience of one Tom Wicker, a white soldier placed in charge of a railroad passenger car with two other white and twenty-seven black soldiers on a train moving troops from Seattle to Virginia in 1946. The trip was to take about two weeks and would conclude with the soldiers' discharge from active duty at its end. Soon after the train started moving, a tall black soldier called out to Wicker, "Hey you, Red!" and immediately a "silence fell on the car like soot from a steam engine" (Kochman, 57). When Wicker replied, "Yeah," the black sailor responded, "Suck my black dick." Wicker found himself suddenly "on the cusp," having to decide how to respond to an obvious provocation. Kochman writes:

> Wicker said half the blacks laughed, a little uncertainly, and one or two eyed him stonily. He could not tell whether he was being teased or challenged. Nonetheless, he was "astonished ... that the tall black thought there was any reason to be hostile, even more astonished that a black man would dare to speak so to a white." He had to respond, but how? He could deal with this black youth as a "Southern white man would deal with a colored person, whether n____, nigruh,

or Negro, and back it up; or else he would have to deal with him as one human with another and live with the consequences." (Kochman, 57)

Wicker decided to risk the latter course and remembered an old joke: "Why, your buddy there told me you didn't have one. Said a hog bit it off." The black soldier grinned in the midst of general guffaws and returned the "cap" with a comment about Wicker's girlfriend. Wicker obliged in kind until another black interrupted "amiably," asking when dinner was going to happen, the train suddenly lurched, and everyone rushed toward the dining car.

That split-second decision to enter into verbal play—risky as it was—was a quintessential instance of verbal *reciprocity* between black and white. It constituted a kind of "turnabout" on black terms, a white person confronted with a black "ritual of insult," forced, on the spot to "declare himself," to make a tiny, but momentous, decision. He could either perform his own identity afresh in that moment, entering into a charged encounter he did not have control of, or he could refuse, fall back on "white" authority, and thus exclude himself from any but a hierarchical relationship, placing himself outside of black culture altogether. The choice to enter into play—really, "to be put into play," himself—was a choice to submit to the other as one human being to another. The initial challenge of the black soldier was a provocation at once expressing the hostility of the situation between white "authority" and black "subordination" and in that very hostility, inviting the white person into an arena of parity where the hierarchy could be leveled. It offered a verbal format that *might* have, as its possible effect, the creation of a certain experience of reciprocity, an exchange of respect in the very act of exchanging calculated putdowns. In any case, Wicker clearly sensed he would have to live with the consequences. But he would have had to do that whichever way he chose to respond.

The Possibility of a Post-White Vocation: Initiation

The other precondition for the lifelong engagement against white supremacy is a necessary *transfiguratio*—a discovery of a different kind of power by learning from people of color and women and other sexual orientations. This involves *initiation*—baptism—under the hands of a wilderness/wildness figure from the margins like John the Baptist. The world of color is not just to be "understood," but undergone. Understanding something implies just that—standing "under" it, giving it power over oneself. More graphically for white males, it means learning how to challenge "white supremacist capitalist patriarchy" not just cognitively, not only politically, not simply in terms of a different circulation of resources and a different cultural habituation, but also as a form of spiritual *incubus*.

White supremacy is one of the preeminent "principalities and powers" of our time. James Cone as we have seen began his life's work by suggesting that black *Power*—not just black freedom—was the message of Jesus Christ to modern America (Cone, 1969, 1–4). One way of reading that claim is to imagine that behind the veil of everyday appearances, at the level where spiritual forces and cultural "climates" compete and war with each other, a partial transfiguration was accomplished by black spirituality and black activism. The "principality of white supremacy," we might say, was wrestled into yielding a different kind of potency; a power of oppression was partially tamed and forced to become a power of creativity.

Saying this is not to say black power is simply a refraction of white power—it remains irreducibly unique and nonderivative as a historical force and a collective impetus. But it is to say that "the master's tools have been regularly used against the master's house" with great resourcefulness and prodigious innovation. Neither is this evaluation to be dismissed as romanticization. Black Power as a movement certainly had its excesses and immaturities, as we saw in chapter 3. It was obviously quickly repressed and bought off (by strategic assassinations of some leaders, imprisonments of others, and co-optations of yet others), marginalized and splintered by the reactions (and partial reforms) of the dominant racist order (Omi and Winant, 84–91).

But at the level of spirit, it is suggestive to entertain the idea that, among its other accomplishments, the Black Power movement represents a kind of "spiritual judo" exercised on white power, in which the violent energies of a dominating force were identified, rebuked, raided, and ritually reconfigured into an aggressive and assertive counterpower that has not ceased to have potent effects in the social order (Omi and Winant, 95–112). Not least of those effects was the unmasking of liberal white color-blindness as itself "predatory" and an insistence, in kind, that specific differences between ethno-cultural groups be recognized and valorized. Not only were black people to be embraced as legal, political, and economic equals, but also recognized as cultural, spiritual, and social innovators of "otherness." That otherness is not and never has been either equitable with or answerable to whiteness.

And it is just this innovated otherness that represents a paradoxical "rebuke and invitation" to white people in general and white males in particular. Black "power" is a resource that white people committed to lifelong antiracist struggle dare not touch . . . and dare not *not* touch. It is not enough—it will not be enough over the long haul—for whites (merely!) to work for equality for people of color. The reality of what "is" in America in our day is a mix of contributions from all kinds of people hidden behind a veneer of white power. If white people are going to labor for a different kind of world, part of the labor that must be undergone is that of being themselves reconstituted in a different form of power. Empowerment for struggle is crucial. But so is the question of whose power and what kind of power. How antiracist whites guard against simply one more time reproducing another realization of the power of whiteness—precisely *in* their antiracist activities—is no mean question. Black power stands out before whiteness as an insoluble conundrum that can also offer irreplaceable reflexivity. White people cannot become black. They can, however—in encounters with others in their spiritual and cultural depths—become more than just white.

The issue is not just awareness, but experimentation with, and *immersion in,* and indeed alteration by, subjugated knowledges (Foucault, 1980, 92–108). Allowing black embodiments of powerful cultural forms to "enter" into one's own sensibility, to come up inside of one's own body with all the terror that implies, is necessary for the breakup of white hegemony and habituation. Mystics have sometimes said "it is not possible to perceive in another reality until one has acted in that other reality." At stake is not just a new cognitive awareness and objectivity about the situation of race, but a new passionate posture and subjectivity founded on a new spiritual interiority. Long-term struggle requires what Brecht once called "a long anger." It is a question of allowing a new historical passion to be engendered within oneself that can orient and motivate lifelong combat *because* it taps into currents of vitality that are larger than one's own resolve to remain faithful. As British cultural critic Paul Gilroy has asserted, it is not enough just to be *anti*racist; there is need to develop more than mere reaction (Gilroy, 1987, 114, 150). Expressive black cultures represent a positive and creative articulation of power (Gilroy, 1987, 156–160). The same question emerges with respect to white antiracist commitment.

What would be required for whites (and especially males) to have formed within themselves the kind of vital ferocity necessary to struggle lifelong against the powers and privileges that whiteness constantly leverages in and around them? The obvious resource is the community that has had no choice about such a struggle. But is it possible for white people to relate to expressions of black power in such a way that they are simultaneously warned away from one more gesture of appropriation *and* initiated into their own powers of antiracist responsibility? It is not a question of wholesale repudiation of cultural borrowing (such is impossible), but rather of sustained self-confrontation and self-exploration, on the one hand, and public acknowledgment and economic return to the black community (commensurate with what was borrowed), on the other. (In one sense, hip-hop culture today is radically "blackening" white youth culture—as indeed youth culture of many ethnic backgrounds—across the globe, but without any accompanying pedagogy that inculcates respect for the history or recognition of the cost of the expressive capacity so developed and repayment to the community that has created it. This I consider to be a failure of the rest of us and thus another part of the unpaid reparations that some day, for better or worse, will come due.)

My own experiences, mentioned in chapter 1 of this writing, have both cautioned and confirmed my attraction to black cultural creativity. Anger, in particular, has become a resource that I value in my work in the classroom as well as in the poetry I perform in the city and around the country—anger not so much splashed out as decanted over years into an energy for change and for discernment of the world around me. Part of my own task of maturation has involved moving past teenage resentment of my father's blustery style of parenting—a reaction that only gradually galvanized into respect as I learned how to embody my anger rather than project it, and use it as a tool for self-knowledge and self-discipline. At the same time, from my thirties on, I became more and more aware of a deep capacity for what I would call a "percussive articulation of indignation and of joy" in certain black friends of mine—a capacity that could run circles around my own nascent attempts to give voice to either *joie de vivre* or injustice. And their gift was not merely *individual* genius (though it was also certainly that).

The more I paid attention to what was around me in inner-city Detroit and later in Chicago, the more I recognized a cultural cadence that lent itself to passionate and incisive commentary, rooted in a seemingly bottomless depth of groan (that I associated with the history and opacity of black suffering). Hip-hop culture today embodies something of that inchoate depth of rhythmic insurgence (even when its lyrics are stupid), as did the Last Poets in the 1970s and Malcolm in 1960s and certain harder-edged modalities of jazz before them (and I am sure they were giving improvisational voice to a competence "schooled" into them by anonymous ordinary folk who had learned from their peers and parents before them). I simply fell in love with the depth and remarkable sinuosity of an intelligence combining precision with passion—and did so, I am sure, because it augured something kindred in myself. After years of living in the context of such richness, I reached a point where I could not *not* risk expressing it in my own voice. Poetry was born in my life, and ever since, performance in the city and animated intensity in the classroom have become my particular venues for continuing to explore that intersection of African American polyrhythmic power and my own personal pain. I wrestle with recurrent moments of anguish that my own expressive artistry cannot (as long as white supremacy remains intact) translate into a simple and clear mode of identification with a historic community of struggle (such as the black community). But I do thrill at what has been opened in me and treasure the partial embrace I (sometimes) receive from black colleagues and strangers—alongside of their ongoing confrontation and challenge to step even more radically away from the lurking supremacies of my white upbringing. I have become a different person-in-living by way of becoming a different body-in-expression and find that inner recuperation of passion essential to my daily struggle against white ignorance and racial injustice.

The Sacrament of a Post-White Vocation: Baptism

The theological model for such a white spiritual transformation is perhaps analogous to Paul's admonition to the Galatians to submit to a reciprocal and collective gestation process, "until Christ be formed again within them" (Gal. 4: 19). Apparently the Christ they thought they had come to know was not the real thing, but a figure and fiction of their own social logic. They had to be "re-incubated" spiritually— forcibly excised from their seductive myth of privilege and pushed back and down into the messy inchoate-ness of a new birth. They had to learn how to "groan" again. In the Pauline vision, human groaning opens out into the groan of all of history and time, indeed the very groan and urge of creation itself against bondage, the primal groan of the Spirit over chaos (Rom. 8: 12–27; Gen. 1: 2). For him, the spirit is a force of creative convulsion, and messianic birth pangs are in evidence everywhere.

What would it take for white "Galatians" today to learn to labor at such a depth? Would it be possible for something like the "agony and ecstasy of Christ inside blackness" to be brought to conception within a white person or community? For white males in particular, "white male pain" is the obvious way into such deeper dimensions of passion incarnate in history. But white male pain is itself a socially produced and mediated sentiment, that feeds on and enforces other forms of socially produced agony (in subjecting women and people of color to its rages and demands, for instance). The problem with much of the "men's movement" in this country, in focusing on the pain of being "male," is that it takes seriously neither "pain" nor "position." Only the top levels of the hierarchy of anguish in our society—only that which is specific to the experience of being a white male—are addressed. And when addressed, solace is quickly sought rather than allowing that pain to become a kind of Dantesque guide to sound out "other" depths.

In the face of such a failure, baptism is an interesting figure. It invites immersion into the death and resurrection of an executed God (as Mark Taylor might say)—but a God profoundly incarnate in history. In American history, that passion, in both of its aspects of "agony" and "overcoming," is perhaps nowhere as profound as in black culture (with the possible exception of Native American culture). Such is the meaning of Cone's claim that Jesus *must be* black in this country. But baptism then implies that all of us, whites included, are in fact *already* being plunged into "blackness" as the very condition of any encounter with the Spirit of justice. To the degree "Christ" is really being formed in any of us as Americans (and not some mythic messiah of our own devising), "black pain and power" are already at work within us. But are whites capable of letting such a "black passion" live in them, without disingenuously trying to disown their whiteness? Or does white baptism, in fact, only occasion deep interior denial and confusion?

The real task of baptism—its profound "initiatory" function—is to plunge one into the depths of the world, below words, where experience is not yet colonized entirely by form, to be remade there, in a new form, not controlled by oneself, not beholden to one's own position, not mapped by one's own social programming. The typical Christian practice here could learn much from older indigenous notions of rites of passage. The need is not momentary, a brief "wetting" under the waters of otherness, but rather a "re-scripting" of the body over a substantial period of time. African bush schools, for instance, might last anywhere from a few weeks to many months, during which time the world of the initiate was inverted, one's mother tongue spoken in reverse ("up" now meaning "down," "left" standing for "right," "hot" for "cold," etc.), the body put through a process of repatterning, lying supine in the dark for long hours, eating strange foods, learning new motions of dance and hunting, facing painful incising of the flesh and fearful purging of the psyche through rituals of terror and healing. For white people the functional equivalent of such a "baptismal" reprogramming is lifelong self-discipline and self-confrontation in the existential schools of racial encounter, inculcating a different habit of perception, able to see and feel the significance of the entire system of supremacy that bears down with such intransigent weight under the skin of advertised equality. Before there can be cognition of a new possibility there is need for deeply felt consternation inside "the problem." The problem is a code of absolute differentiation habituated deep inside the white body that requires sustained confrontation and interior work.

The interior effect of such baptismal work would be threefold. On the one hand, it would involve a disinterring of the *theological structure of whiteness* (under its secular modernization) as a mode of identity ever demanding an absolute reference point ("blackness" or other perceptions of "color") against which to feel secure. "Wholeness" for whiteness has ever been structured as a binary "salvation schema" anchored in a drive for certainty. Healing from supremacy, at the level of embodiment, would mean immersion in uncertainty, letting go of the body as a fixed coordinate in a map of meaning, relinquishing integrity as something controlled by the clarity of a binary structure of "this, not that." Opening to salvation as a process, to exchange with others as its means, to grace as something that comes from outside one's identity structure, to spirit as an energy that "breaks out" within, requires abandonment of the dogmatic reflex that locates identity and wholeness entirely on one side of an opposition. Only so, can deep understanding of the situation of race begin to etch itself inside white awareness and begin to create, as its second moment, capacity for a new structure of passion, a new "long anger" capable of incubating and animating a will-to-change over the course of a lifetime. And finally, such a wrenching reorganization of habitual denial and emotional presumption would entail a kind of reworking of the connection between cognition and conviction, a new linkage of perception and passion, giving sharp focus to profound feeling, making of the tongue a weapon capable of "carving meat from bone,"[9] of outing supremacy wherever it lurked in the wings of white power.

Baptism understood in such an initiatory framework is simultaneously a plunge "outside the body" into the life experience and cultural resistance of those the social order regularly "rolls over." As Asian theologian Aloyius Pieris has argued, baptism is double (Pieris, 62–63). As exemplified in the life of Jesus, it began in a river and culminated on a hill (Lk. 3: 15–22, 12: 49–53). For him, it was not just a matter of going under the waters and coming up under the dove, but of placing himself *in* the hands of another prophetic leader and of committing himself *to* a social movement already in progress. His immersion designated a lifestyle and exhibited solidarity. It amounted to a choice about whom he would be willing to die with and what he would die for. Jordan led to Calvary.

But it did so by way of Capernaum. Jesus did not simply "study" the ethnic and economic and gender wisdom (i.e., the black and liberation and feminist "folk theologies") of the peasants and day laborers and bent-over synagogue-goers and uppity sufferers of blood-flows of his day. (Although clearly he did deeply imbibe the street-smarts and arts of resistance of the poor and oppressed he lived among, and regularly offered their actions as examples of faithfulness when instructing his own disciples; see Mk. 7: 24–30; Lk. 9: 25–37, 16: 1–8, 18: 1–8, 18: 9–14.) Rather he entered into their social circumstance, learned from their cultural experience, and, in challenging their oppressors (his own "peer" teachers), embraced their political destiny. He allowed his body to be "occupied by" their energy (the authorities, after all, thought he was "possessed"; Mk. 3: 20–27; Jh. 8: 48). His public discourse was charged with their private anguish (or just how *do* we imagine the emotional "tone" of his public cursing of the leadership elites?; Lk. 6: 20–26; Mt. 23: 13–29). His eloquence was informed with their idiom ("not as the scribes spoke," it says; Mk. 1: 22). Jesus's admonition to Nicodemus was not something he had not himself undergone (Jh. 3: 1–5). The "social immersion" he anticipated by way of water, he lived out in his flesh, celebrated in his partying, relied on in his politics, and grieved the loss of on the night before the bloodletting (Mk. 2: 31–35; Lk. 15: 1–2; Mk. 1: 18, 32, 12: 12, 37; Lk. 22: 15–16).

Baptism, in this frame, is a plunge not only into the waters of life and death in the abstract, but also into the social and existential experience of those around us for whom death is most precipitous and life most precarious. It places the issue of faith on the political and cultural map: it is a matter of allowing one's own experience of embodiment to be accosted by those whose bodies are under assault. It offers "wholeness" in the key of concreteness. We are "drowned" in the anguish of the oppressed of our own time and raised in the powers of their particular forms of resistance. Anything less is docetism (Cone, 1975, 36–37). If we refuse the social reality, we lose the spiritual efficacy. Until and unless we risk baptism in its most profound political and cultural implications in our own historical setting, we render its eternal possibilities illusory and ourselves impotent. In America at the millennium, black embodiment of pain and power remains a primary litmus test of white "baptism." Most of us who are pale are hardly even wet yet.

Conclusion

For me, the pivotal issue in white theology's relationship to black and womanist theologies is this latter question of spiritual and emotional empowerment. I regularly drink from those wells not simply as a matter of according them their "due" hearing. I plunge in rather to rekindle my own out-rage and restrengthen my own *cour*-rage so that I do not fail to take up my own part in the lifetime of social and spiritual struggle that I believe is the only future worth living for. I grow more and more aware, as the years pass, that the witness to what is indomitable in the human spirit is worldwide, multireligious, polyvocal, many-colored and ever-surprising. Even as I discern principalities and powers organizing anew in almost unthinkably monstrous modalities today, I also see resistance on the scale of human-sized gestures of joy and defiance proliferating their own arts. NAFTA conspiracy is met with Chiapas determination; Republican attempts to reorganize affirmative action in Houston provoke a coalition reorganization of politics electing a black mayor. As has ever been the case between domination and its rebuke, it seems to be a race never quite won . . . or lost (Scott, xi).

I personally get knocked down often and depressed readily in my own small sphere of confrontation. But I have learned how to get up, again and again, from very particular friends and very specific communities. They are mostly darker-skinned than I. I take that to be an indication of the other half of the name that this culture still tries to name as "problem" (Du Bois, 1961, 15, 20). That other name might just be something like "inspiration" or even "salvation." If we ever got to a point in this world where white people could embrace a black messiah, we could then perhaps finally agree to consign those two fictions about color to the dustbin of history. Until then, my hope will continue to take its orientation from what comes like a thief, cloaked in night.

Epilogue

In the Genesis story, Jacob is saved only by virtue of having passed through his own ultimate horror. Having stolen his brother's birthright by a ruse of skin, having introduced the fear of death into that relationship by making it subservient to an economic end (inheritance of the family wealth), having usurped power and fled the place of encounter, Jacob finally comes to the end of his own flight and decides to return home. At the crossover point (the Jabbuk river) into Esau's turf, he seeks first to buy reconciliation and then opts to face the entire history of his dishonesty alone at the border. The border is understood to be a place of haunting, an intersection of political power and supra-political principality. He is moving from one dominion to another. There are dues to be paid, a right (and rite) of passage to be negotiated. The exact force that resists him remains nameless, even though the import of the midnight encounter is clear. The terror is one of Dying—in its absolute meaning not merely of physical death, but of that uncanny "other" aspect of death that has something of horror about it. The terror is one of transgressing a certain Ultimacy and facing its face without mediation. Seeing the face of "God," absent a veil of covering, in Jacob's culture, is popularly imagined as a mortal encounter. God here is just as likely Death as Life, Fear as Freedom. The moment is that of the *Tremendum*.

But something in the ultimacy of the encounter *is* freeing. Pretense drops. Disguise disappears. Jacob takes back his projections, ceases pretending to be other than grasping, owns his usurpation and striving and strives with a vengeance. Here, at last, there is no imitation of the other; there is sheer, desperate wrestling. Fighting to the death to name! Fighting to name the Death!

Jacob loses and wins in his losing. He is broken at the point of his greatest physical power, broken in his body, as the mark of daring to face and fight the spirit. This is very like the way Long speaks of slave encounters with the *Tremendum*, an immediate engagement with the unnamable Other whose power to destroy is not merely physical, but spiritual and absolute. Is this "God?" The angel of God? "Demon?" Or all of the above? It matters not; what matters is the risk of power at ground zero of the self, where subjectivity erupts from a mystery that is both alluring and terrifying, irrupts as a form of contingent being that is incalculable. Jacob cannot name his striving or its power. It rather names him as itself. He now is Striving, is Power, is "Israel"—in the body of a human that bears the wound of that encounter without excuse or surrogate. Is this not the historical specificity of the Pauline paradigm of death and resurrection in Jacob's own time and place?

The result is a human who limps the reality of his (now) hybrid history as a hyphenated personality. Jacob the heel-grasper has become Jacob-Israel the god-striver. White in skin, he has become "red" in person, the very potency of the hunter he never became and Esau always was. Ceasing to use his brother as surrogate to face the father he fears and desires, Jacob becomes an Israel that embraces all three. His "doubling" is finally a multiplication of desires and powers. Salvation here is a matter of becoming more than one, of being fractured into wholeness. A healed heart demands a hollowed hip. Jacob-Israel harbors divine power in its only possible human form. The one in the other *is* grotesque.

And such then is the theological backlighting to racial "conversion" in this time and place. The slicing through of joint and marrow by the Word of discernment necessary to effect white redemption is a depth-operation. What for blacks is an irony of agency, a self-re-creation *against* the creation of oneself by an oppressive other, must be undergone by whites as an agency of irony. White agency here is finally an agency against *oneself*, wrestling in a form that turns around on one's own substance and creates two in the place of one. It is perhaps strange to think of wholeness offering itself in this form, but given the positioning of white "being" inside the principality of white oppression, it cannot be otherwise. There is no clean and easy exit from white power in this country. There is only the pain and joy of continual combat, and the work of that combat in carving out a region of wisdom or "soul" inside the ever signified upon white body. Until discourse and practice no longer manipulate light appearance into privilege and position at the expense of dark appearance, "white" skin will remain caught in a demonic web. Only the effort to make it simultaneously bear another meaning—both for oneself and for people of color, both existentially and politically—can tear the web. The effort is the salvation. Even though it is given from without.

Notes

Part I White Privilege and Black Power

1. In addition to the discussion in the last chapter of Riggins Earl, Jr.'s *Dark Symbols, Obscure Signs: God, Self, and Community in the Slave Mind*, cf. also Albert Raboteau's *Slave Religion: The "Invisible Institution" in the Antebellum South*.

I White Boy in the Ghetto

1. See bibliography for book titles by those names; Bynum, 92; Morrison, 1970, 68; Hopkins, 1993, 57–60; Hurston, 197; Porterfield, 728–729, Smith, 1989, 384–387; West, 1989b, 93.
2. In the sense of Nietzsche's "sounding out" of idols and of African American vernacular traditions of "sounding out" and "signifying" (Nietzsche, 1990, 31–32; Gates, 1988, 81, 94).

2 The Crisis of Race in the New Millennium

1. After the song title by that name of the hip hop group, Public Enemy.
2. The African American teenager shot and killed by a Korean shopkeeper on video tape while trying to purchase a pop two months before the announcement of the verdict on the police officers accused of beating King. The shopkeeper reportedly suspected her of stealing and feared attack in the confrontation.
3. For example, *Racial Formation in the United State: From the 1960s to the 1990s*, Michael Omi and Howard Winant, 69, 95–96, 101.
4. Cf. Perry Miller's *Errand Into the Wilderness*.
5. At least up to this point in our collective history. It is theoretically conceivable that at some point in the future, the demographic balance of power could shift in favor of groups that currently remain minority and with it, the possibility of new forms of racism promulgated by such new majority groups. But for the

foreseeable future, racism remains the coercive collective effect of white power structures and white cultural norms.

6. This, in fact, is exactly how James Cone describes white racism: "According to the New Testament, these powers can get hold of a man's total being and can control his life to such a degree that he is incapable of distinguishing himself from the alien power. This seems to be what has happened to white racism in America. It is a part of the spirit of the age, the ethos of the culture, so embedded in the social, economic, and political structure that white society is incapable of knowing its destructive nature" (Cone, 1969, 41).

7. Even when, in a book like *Black Theology & Black Power*, he expressly indicates he is not writing chiefly for black people, but is addressing "a word to the oppressor, a word to Whitey," Cone nonetheless qualifies his motive by immediately adding, "not in hope that he will listen (after King's death who can hope?) but in the expectation that *my own existence* will be clarified" (Cone, 1969, 3; emphasis added).

8. This might appear to be a question typologically similar to Rosemary Radford Ruether's feminist question, voiced as the title of a chapter, "Can a Male Savior Save Women?" in *Sexism and God Talk*. But it is a deceptive similarity, as Ruether's question is about the relevance of a dominant culture construction for those in the position of the oppressed. Here the question is inverted.

9. Susan Thistlethwaite, for instance, deals with race and gender together in a compelling manner in *Sex, Race, and God: A Christian Feminism in Black and White*.

10. Witvliet traces the effects of this dynamism in both Martin Luther King's response to Vietnam, and Malcolm X's to Mecca. Each began by speaking for and to a *particular* constituency, but ended by pursing a *global* vision.

11. Cf. Henry Louis Gates, Jr., *The Signifying Monkey: A Theory of African-American Literary Criticism*, chapter 2.

12. See Pierre Bourdieu for the notion of *habitus*, the enculturated and internalized patterns of perception and calculations of response to one's cultural environment and social others that normally are simply taken for granted and thus operate almost outside of consciousness, but which can be brought to the forefront of intentionality if put under pressure by sudden social change or contact with another cultural formation (Bourdieu, 17).

13. For another expression of this same point, see Karen McCarthy Brown's description (in *Mama Lola*) of how possession performances in a Haitian *Vodun* community in Brooklyn facilitate experimentation with various forms of social conflict faced in "real life" and the various feeling-structures and bodily postures required in different modes of response.

Part II   History, Consciousness, and Performance

1. And thus theologically "legitimate" in the white academy (in some quarters at least).

2. Cf. especially, Thomas F. Slaughter, Jr.'s "Epidermalizing the World: A Basic Mode of Being Black," in *Philosophy Born of Struggle: Anthology of Afro-American Philosophy from 1917.*

3   Modern White Supremacy and Western Christian Soteriology

1. For an account of this New Testament idea in relationship to more modern notions of various forms of collective power, see in particular, Walter Wink's trilogy on the Powers and William Stringfellow's *An Ethic for Christians and Other Aliens in a Strange Land.*

2. I am indebted, for the basic insight here, to George Pickering in a talk he gave on April 24, 1995, at the University of Chicago Divinity School during a conference entitled, "'Our God is Able': A Retrospective on the Civil Rights Movement as an Interfaith and Ecumenical Movement."

3. Cf. Dwight Hopkins, in *Shoes that Fit Our Feet*, where he says, "If anything, the white church's systematic lineage and practice of racism call forth the Devil and certainly do not reveal a just God" (Hopkins, 1993, 144).

4. Cf. especially T. Todorov, *The Conquest of America* and V. Mudimbe, *The Invention of Africa*. It is also the case that indigenous soteriologies have figured quite dramatically in native receptions of Europeans, such as in the Hopi Indian struggle to "place" Coronado's dispatch, Pedro de Tovar, in the Sixteenth century, or all the white settlers that have followed since, in their own redeemer-myth of the return of Pahana, the White Brother who will bring about the transition of this world into its next phase (Rudolf Kaiser, *The Voice of the Great Spirit: Prophecies of the Hopi Indian*). Cf. also Aztec mis-figurations of Cortez as the promised Quetzalcoatl as detailed in Dussel's *The Invention of the Americas.*

5. Cf. Jose Rabasa, *Inventing America: Spanish Historiography and the Formation of Eurocentrism.*

6. The Wild Man or Woman represented the distillation of anxiety about the three main areas of life supposedly secured and civilized in the Christian institutions of family, society, and church: sex, sustenance, and salvation.

7. Cf., e.g., Mudimbe, who traces the way notions of blackness filter back into European artistic traditions and religious explanations of cultural difference (Mudimbe, 7–9).

8. Cf. Frantz Fanon, commenting on George Balandier in *Black Skins, White Masks* (Fanon, 95).

9. Cf., e.g., the debates recounted in Dussel's brief summary of the Valladolid disputes between Bartolemeo de las Casas and Sepulveda in sixteenth-century Spain (Dussel, 1995, 64, 67).

10. I have chosen to use the term "America" in this writing in its popular rather than technical sense. Technically, the United States is only a part of America, which actually encompasses all of the northern, central, and southern land masses and peoples that began to be referred to as the "Americas" some time

after Amerigo Vespucci's name was first appended to the geography Columbus had mistaken as India.

11. Bastide's discussion is not limited only to Protestant configurations of race. He argues that color symbolism articulates with social practice differently in contexts dominated either by Protestantism and Roman Catholicism, tending to show up in the latter not so much in severe strictures of segregation between black and white as in various forms of color hierarchy that govern "racial interbreeding" (Bastide, 276).

12. Bastide notes the Christian justification of slavery in which the claim was made that "black skin was a punishment from God" (Bastide, 272).

13. Bastide argues, e.g., that although black and white "have taken on other meanings," the "'frontier-complex' between two conflicting mentalities has held firm" (Bastide, 285). And these other meanings then "still follow . . . the basic antithesis founded centuries before on the white purity of the elect and the blackness of Satan."

14. In tracing the ramifications of the black/white symbolism through "the double process of secularization in America and of de-Christianization in Europe," Bastide claims, that in America, "Calvinism remained just under the surface, ready to be revived at the slightest opportunity" (Bastide, 282).

15. For a theoretically dense rendition of this picture of modernity, see Bhabha's "Dissemination," in *The Location of Culture*, 139–170.

16. This is the burden of Dussel's argument in *The Invention of the Americas*.

17. Cf. Steven Ozment's *The Reformation in Medieval Perspective*.

18. Hegel carries the fracture begun with Kant, between self as subject and self as object, to full term in the final reconciliation of identity and difference as an identity between "identity" and "identity-and-difference" dialectically achieved in Absolute Spirit. Marx, in turn, stands Hegel on his head, identifies the universal class as the proletariat, within whose coming into full possession of the forces of production (in communism) will be accomplished the resolution of historical contradictions in the classless society.

19. The idea that "like can only be saved by like" has a long history in Christian theology, going all the way back to the patristic era. In *Inventing Africa*, Mudimbe traces some of the changing effects of this way of thinking by relating the Foucautian model of epistemic shifts to the colonizing process (Mudimbe, 9).

20. Cf. Max Weber, *The Protestant Ethic and the Spirit of Capitalism*.

21. Cf. Richard Kearney's *Transitions*, tracing the struggle of the bourgeois hero of the novel to invest the world of modern meaninglessness with some intimation of transcendence absent input of a providential god (Kearney, 35).

22. Including contemporary efforts such as Andrew Murray's attempt to re-racialize "intelligence" in *The Bell Curve* (Herrnstein and Murray, *The Bell Curve*), or American Association of Criminology President James Q. Wilson's theories of "somatotyping" (addressed in chapter 6).

23. Cf. Michel Foucault's *The Order of Things* for one accounting of the shift.

24. In one sense, Weber's entire project in *Economy and Society* is an attempt to account for this distinctive emergence of spheres of autonomous interests and knowledges in Western modernity as compared with other social formations developed elsewhere in the world.

25. The "revised Marxisms" in question would include those issuing from the Birmingham Center for Cultural Studies connected with names like Stuart Hall and Dick Hebdige, and those showing up among various colonial and postcolonial discourse theorists like Gayatri Spivak (Spivak, e.g., has described herself as a feminist, deconstructionist, Marxist; Spivak, 1990, 104, 116).

26. For further discussion of this issue see Juan Luis Segundo, *Faith and Ideologies*; Clodovis Boff, *Theory and Praxis*; Claude Lefort, *The Political Forms of Modern Society*; and Ernesto Laclau and Claude Mouffe, *Hegemony and Socialist Strategy*.

27. Cf. Gilroy's argument in the following section.

28. Spivak, e.g., identifies Kant's critique of Descartes as the inaugural moment in the "breaking up of the individual subject" of modernity (Spivak, 1988, 310).

29. For instance, Hume's nature as "divine mind" shriven under Kant's critical acumen.

30. As we shall examine in some depth in chapter 4.

31. Literally the "other" who is "under," a class of those whose voice and agency are submerged in a hegemonic social formation. Cf.Spivak's "Can the Subaltern Speak?" and Benita Perry's critique (Spivak, 1988, 271–313; Perry, 35).

32. An incident that has found new currency in a reinvented form in Toni Morrison's *Beloved*.

33. Cf. Alice Stone Blackwell, *Lucy Stone: Pioneer of Women's Rights*, 183–184.

34. Cf. Anna Bontemps, *Free at Last: The Life of Frederick Douglas*, 180.

35. It is worth noting, here, although it is beyond our exploring in depth, that Jamaican Rastafari have developed an implicit "theology of dread" that informs their religious practice. "Dread" becomes, in part, something they associate with their own experience of "Babylon," the white Euro-world-structures that enslaved Africans and now oppresses their diaspora descendants. Babylon dominates for the present, but faces in the future a "dread" judgment and downfall. "Dread" locks, the long natural curls worn by Rastafari men and women, symbolize a certain "leonine" presence that is also embodied at times in their silent defiant presence in white society, reflecting both their own certainty of coming judgment and a kind of congealed opaque testament of all of the dead who have suffered Babylon in the past and now await vindication. "Embodied dread" is thus significant of living power for black sufferers and of an impenetrable apocalyptic exclusion of white oppressors. It is the sign of the irresistible end of exile and a return to the glories of "Ithiopia" (Africa) that requires no comment from blacks and brooks no response from white Babylon. I am indebted for these comments to anthropologist Greg Downey from the University of Chicago in a conversation we had in 1995.

36. Cf. Theophilus Smith's Girardian reading of the Civil Rights movement, in which the movement's strategy is analyzed as the continual staging of a kind of public inoculation ritual for North American racial violence by way of the (black led) "*pharmakon*" of nonviolence (Smith, 1994, 211–218). The trick was to create the conditions for enough violence to manifest itself to

be undeniable and inexcusable, and simultaneously to build in enough controls on the situation to avoid any absolute victimage.

37. Cf. Vincent Harding's 1967 article "Black Power and the American Christ" giving provocative articulation to Black Power as a spiritual counterforce in part brought into being by the blue-eyed, blond-haired "American Christ" of white Christianity in North America (Harding, 40).

4   Black Double-Consciousness and White Double Takes

1. Cf. James Scott's notion of hidden transcripts that are part of the in-group consciousness and conversations on both sides of the divide between domination and subordination, normally camouflaged and only under pressure erupting in contravention of the "public transcript" tacitly agreed to and maintained by both sides (though imposed and policed by the dominant). In reference to white supremacy, the idea of a white "hidden transcript" fits well with Charles Mills's notion of the Racial Contract (Scott, 14; Mills, 11).

2. I am indebted for some of my thinking on Du Bois in what follows to a presentation of a paper by anthropologist Nahum Chandler in a Workshop on Race and the Reproduction of Ideologies at the University of Chicago, May 2, 1992 as well as two subsequent articles in *Displacement, Diaspora, and Geographies of Identity* and *Callaloo*, respectively (Chandler, 1996a, 240, 250, 253; 1996b, 83–84, 86, 90, ft. nt. 7).

3. In dealing with a different difference, Freud, too, will halt before the "split" of this kind of "split-second" look, catch it, give it extension in a piled-up Germanism, rendering it unlike itself as an *Augenblick*, a seeing washed by lids, but not washed away.

4. Chandler says, e.g., that "although critical of the indecisiveness and incoherence that this sense produced in Negro political and social life, private and public, and deeply responsive to the *violence* of the *sense* of this heterogeneity, the actual experience of this sense, Du Bois never ceased to affirm this heterogeneity as *also* a good, a resource, in general" (Chandler, 1996b, 85).

5. Part of the seminal originality of Du Bois's work, according to Chandler, is the fact that DuBois quite carefully thought through his experience of race not so much as an isolated individual, but rather as one inscribed within a system, written upon and partially ruled by, a "concept," whose career in his own life history, Du Bois then tracks and rewrites. Du Bois's own description of what he is doing, underscored as the subtitle of *Dusk of Dawn*, is working out an "autobiography of a race concept" (Chandler, 1996a, 240–250).

6. As in the trope on Marx by historian Thomas Holt to the effect that, "human beings make race, but they are not free to make it in any old way they please" (offered in a comment during the Newberry Seminar in American Social History, University of Chicago, May 30, 1991).

5   Black Performance

1. It is impossible here not to note the most recent reiteration of this entire equation—the video-taped beating to the death of Nathaniel Jones on November 30, 2003, by four white and one black officers of a Cincinnati police force that had only in April of 2001 provoked a full scale "riot" for shooting a black youth in the back who was fleeing arrest. The police were summoned to deal with Jones because he suddenly began dancing in some privately conceived revelry in a fast food restaurant.

2. As quoted in the John McDermott edited, *The Writings of William James*, 3 (from Henry James, Sr.1879.*Society: The Redeemed Form of Man*. Cambridge: Hougton, *Osgood & Co.*, 44–49).

3. Remarkably from an American theological point of view, what Long labors into expression here by a way of a Lutheran (Ottonian) category could be comprehended as a failure of Calvinism. The *majestas* of Divine Freedom finds its proximate human symbolization not in the Puritanism of the American founders, but among the *sufferers of* Puritan theological foundationalism. In one sense, we could say "sovereign majesty" becomes accessible, as the divine reality that "overshadows" the "total depravity" of humanity, only in the concreteness of history. It is not an effect of theological formulation, but of historical experience—the experience of those who have been made to know depravity "in the flesh." The foundation of the nation—the Puritan covenant made material in the Jeffersonian "slave" republic—is the hard matter against which *majestas* splintered into a *tremendum* horrified at its own reduction to mere fascination.

4. I am indebted for this idea to a comment of Long delivered at the 2003 American Academy of Religions conference.

5. The choice of the word "tactical" here is informed by Michel de Certeau's way of differentiating *tactical* uses of spaces and practices, by those who are not invested with institutional powers and authority but must seize opportunities to resist order "on the run" as it were, from those who do enjoy institutional places from which to mobilize *strategies* of domination (de Certeau, xiv, xix, 35–39). I use the word "strategies" for white practices of whiteness in chapter 6.

6. A play on the Buddhist notion of the "Third Eye," the "wisdom eye" that is opened in meditation practice.

7. Cf. Paul Connerton's exploration of the role of the body in social memory in his *How Societies Remember*.

8. Saying such is already a dangerous statement, obviously vulnerable to a de-humanizing tendency that simultaneously bestializes and divinizes, but refuses the humanity of the other as simply that: ordinary humanness. At the same time, however, it is obvious that the problem of otherness does not just go away because we decide to call it "merely" human. We will return to this question in chapter 8. The place of difference between human beings remains one of the questions that must be attended to with theological diligence and critical acumen. Its potency as a source of theophany or sacral encounter will not be exorcised of violence by mere fiat or denial. At stake rather is the kind of sacrality

that is attributed to the otherness that appears there, the way a theological claim is made for a new possibility of human wholeness and salvation, whether divinity is linked to the exclusionary terrors of a scapegoating operation or the honoring of all human flesh as incarnational. Cf. Rene Girard, *Violence and the Sacred*, and a critical commentary on Girard by Tod Swanson, "Colonial Violence and Inca Analogies to Christianity," *Curing Violence*.

Part III   Presumption, Initiation, and Practice

1. Cf. the citation by James Cone of a Malcolm X speech in which Malcolm asserted that by and large his castigation of the white man as a "devil" had reference primarily not to any individual white man, but "the *collective* white man's *historical* record . . . [of] cruelties, and evils, and greeds, that have seen him *act* like a devil toward the non-white man" (Cone, 1991, 103; emphasis added).

6   White Posture

1. In many urban centers, "profiling" results in the arrest of many poor young males of color with virtual impunity, who are then encouraged to plea bargain for a reduced sentence rather than fight for their innocence championed by an overworked, underpaid, often undereducated court-assigned attorney in a court room under pressure to clear its dockets as rapidly as possible.
2. Cf. footnote on section entitled Tactical Performances in chapter 4.
3. As the gospel of Matthew might have it (Mt. 12: 43–45).
4. As some of the recent literature on the Christian men's movement, Promise Keepers, e.g., details—"the late 18th century transition from a masculine identity based on community and family service to one based on individual conquest in the capitalist marketplace" has bequeathed the twentieth century its stereotypes of maleness as "tough, stoic, isolated, unemotional," constituted in competitiveness, freely expressing the "manly passions" of assertiveness, ambition, avarice, lust for power" (Deardorff, 77; Cole, 125).
5. In the words of philosopher Thomas Slaughter, Jr.—giving phenomenological articulation to the predicament of blackness in a white world "which is the omnipresent possessor of all goods, save some exotic African artifacts"—the "act of pigmentizing the world extends even to the natural world. Today all grains are hybrid; all grass has been sodded; all trees have been seeded; all rivers dammed. And throughout, the producer's imprint is White" (Slaughter, 285).
6. A reference to Frantz Fanon's characterization, in the words of Slaughter, of "contemporary cultural contact between a dominating Western world and colored, colonized and neo-colonized people the world over. He thereby invokes the ancient philosophy which conceived the cosmos in terms of a struggle

between the principles of Good and Evil. In that scheme, black was designated the color of evil, and White, the color of 'Right' " (Slaughter, 284).

7. A sense that is reinforced in blacks if there is also present, in the same classroom, some amount of "countercultural" ("lower class") white input in a style more closely resembling black argumentative style (Kochman, 30).

8. Of no small interest here is the Spanish colonial term "reducciones" for indigenous peoples "reduced" from their wildness to more civilized manners of dress, speaking, believing, and living.

7   White Passage and Black Pedagogy

1. Mills surveys the estimates (conservative and liberal) to arrive at a figure of roughly 100 million killed by European colonial violence directed at nonwhite populations around the world—easily the most heinous case of genocide in history (Mills, 98, 142, ft. 35, 146, nt. 21).

2. Compare the Apostle Paul's descriptions in I Cor. 4: 9–13; II Cor. 11: 23–29.

8   Anti-Supremacist Solidarity and Post-White Practice

1. Cited in a talk given by Joe Feagin, entitled "Racism and the Coming White Minority," on April 20, 2000 at Wayne State University, Detroit, MI.; cf. also Mills, 38–39.

2. Cf. the work by Nibs Stroupe and Inez Fleming entitled *While We Run This Race: Confronting the Power of Racism in a Southern Church* for a metaphorical discussion of race as "infection."

3. Cf. Long's argument for such in *The History of Religions: Retrospect and Prospect*, ed. Kitigawa, 87–104.

4. Parts of this section and the one following entitled "White (Anti-)Theology and Post-White Practice," appeared (in different versions) in "Rage With a Purpose, Weep Without Regret: A White Theology of Solidarity," *Soundings* Vol. 82, Nos. 3–4 (Fall/Winter 1999), 437–463; "White Church Response to Black Theology: Like a Thief in the Night," *Theology Today*, Vol. 60, No. 4 (January 2004), 508–524; and "Black Theology and the White Church in the Third Millennium: Like A Thief in the Night," in *Living Stones in the Household of God: The Legacy and Future of Black Theology*, ed. L. Thomas, Minneapolis: Augsburg Fortress, 83–98.

5. Though as remains abundantly clear in continuing episodes of "profiling" and arrest, it does continue to draw attention and surveillance, or worse.

6. Like the beating of Rodney King seemingly "verified" his supposed violence in the eyes of the Simi Valley jury that acquitted the police officers involved (he was being hit by officers of the law, therefore he must have been a threat).

7. Suburbs, and the highways leading to suburbs, e.g., were constructed with Federal monies and backing (in the mortgage guarantee corporation) supplied

by tax dollars collected from everyone but generally made available, from immediately after World War II until the mid-1960s civil rights legislation, only to white identified people. The Federal budget, in effect, functioned as a transfer payment out of communities of color and into white community wherewithal.

8. Gloria Albrecht discusses something similar in her white feminist work on ethics, *The Character of Our Communities*, 95, 98.

9. In the ritual metaphor of the author of Hebrews, concerning the operation of the word of God in revealing intentions and opening interiority to a new saving scrutiny (Heb. 4: 11–13).

Bibliography

Albrecht, Gloria. 1995. *The Character of Our Communities: Toward An Ethic of Liberation for the Church*. Nashville: Abingdon Press.

Alighieri, Dante. 1970–1976. *The Divine Comedy*. Trans. with commentary by Charles S. Singleton, 3 vols. Princeton: Princeton University Press.

Anderson, Victor. 1995. *Beyond Ontological Blackness: An Essay on African American Religious and Cultural Criticism*. New York: Continuum.

Augustine. 1960. *The Confessions of Augustine*. Trans. by J. K. Ryan. Garden City, New York: Doubleday & Company, Inc.

Baker, Houston. 1984. *Blues, Ideology and Afro-American Literature*. Chicago: University of Chicago Press.

———. 1993. "Scene . . . Not Heard." In *Reading Rodney King, Reading Urban Uprising*. Ed. R. Gooding-Williams. New York: Routledge, 38–50.

Bakhtin, Mikhail. 1984. *Rabelais and His World*. Trans. H. Iswolsky. Bloomington: Indiana University Press.

Baldwin, James. 1963. *The Fire Next time*. New York: Dial Press.

———. 1993. *The Fire Next Time*. New York: Vintage International Books.

Bastide, Roger. 1970. "Color, Racism, and Christianity." In *White Racism: Its History, Pathology and Practice*. Ed. B. N. Schwartz and R. Disch. New York: Dell Publishing Co., 270–285.

Bellah, Robert. 1985. *Habits of the Heart*. Berkeley: University of California.

Bhabha, Homi. 1994. *The Location of Culture*. New York: Routledge.

Blackwell, Alice Stone. 1930. *Lucy Stone: Pioneer of Women's Rights*. Boston: Little, Brown.

Boff, Clodovis. 1987. *Theology and Praxis: Epistemological Foundations*. Trans. R. R. Barr. Maryknoll: Orbis Books.

Bontemps, Anna. 1971. *Free at Last: The Life of Frederick Douglas*. New York: Dodd, Mead.

Bourdieu, Pierre. 1977. *Outline of a Theory of Practice*. Trans. R. Nice. Cambridge and New York: Cambridge University Press.

Brown, Karen McCarthy. 1991. *Mama Lola: A Vodou Priestess in Brooklyn*. Berkeley: University of California Press.

Burke, Kenneth. 1969. *A Grammar of Motives*. Berkeley: University of California Press.

Butler, Judith. 1993. "Endangered/Endangering: Schematic Racism and White Paranoia." In *Reading Rodney King, Reading Urban Uprising*. Ed. R. Gooding-Williams. New York: Routledge, 15–22.

Bynum, Edward Bruce. 1999. *The African Unconscious: Roots of Ancient Mysticism and Modern Psychology*. New York: Teachers College Press.

Chandler, Nahum. 1993. "Between." *Assemblage* 20: 26–27.

———. 1996a. "The Figure of the X: An Elaboration of the Du Boisian Autobiographical Example." In *Displacement, Diaspora, and Geographies of Identity*. Ed. S. Lavie and T. Swedenburg. Durham: Duke University Press, 235–272.

———. 1996b. "The Economy of Desedimentation: W. E. B. DuBois and the Discourses of the Negro." *Callaloo* 19/1: 78–93.

Cleage, Albert. 1968. *The Black Messiah*. New York: Sheed and Ward.

Cleaver, Richard. 1996. *Know My Name: A Gay Liberation Theology*. Westminster John Knox.

Cole, Robert A. 2000, "Promising to Be a Man: Promsie Keepers and the Organizational Constitution of Masculinity." In *The Promise Keepers: Essays on Masculinity and Christianity*. Ed. Dane S. Claussen. Jefferson, NC: McFarland & Co., Inc., 113–132.

Coleman, Wanda. 1993. "Primal Orb Density." In *Lure and Loathing: Essays on Race, Identity, and the Ambivalence of Assimilation*. Ed. Gerald Early. New York: Penguin Books, 207–226.

Cone, James. 1969. *Black Theology and Black Power*. New York: Seabury Press, 1969; twentieth anniversary reprint, San Francisco: Harper & Row, 1989.

———. 1975. *God of the Oppressed*. New York: Seabury Press.

———. 1991. *Martin and Malcolm and America: A Dream or A Nightmare*. Maryknoll: Orbis Books.

Connerton, Paul. 1989. *How Societies Remember*. New York: Cambridge University Press.

Conrad, Joseph. 1996. *Heart of Darkness*. Ed. R. C. Murfin. Boston: Bedford Books of St. Martin's Press.

Crenshaw, Kimberlè and Peller, Gary. 1993. "Rell Time/Real Justice." In *Reading Rodney King, Reading Urban Uprising*. Ed. R. Gooding-Williams. New York: Routledge, 56–70.

Daly, Mary. 1973. *Beyond God the Father*. Boston: Beacon Press.

Dante, Alighieri. 1954. *The Inferno: Dante's Immortal Drama of a Journey Through Hell*. Trans. J. Ciardi. New York: Mentor Books, the New American Library, Inc.

Davis, Mike. 1992. "Fortress Los Angeles: The Militarization of Urban Space." In *Variations on a Theme Park: The New American City and the End of Public Space*. Ed. by M. Sorkin. New York: The Noonday Press, 169–181.

———. 1993. *The War Against the Cities*. London: Verso.

Deardorff, Don. 2000. "Sacred Male Space: The Promise Keepers as a Community of Resistance." In *The Promise Keepers: Essays on Masculinity and Christianity*. Ed. Dane S. Claussen. Jefferson, NC: McFarland & Co., Inc., 76–90.

de Certeau, Michel. 1984. *The Practice of Everyday Life*. Trans. S. F. Rendall. Berkeley: University of California Press.

Deloria, Vine. 1970. *Custer Died for Your Sins*. New York: Avon Books.

———. 1972. *We Talk, You Listen*. New York: Dell Press.

———. 1973. *God is Red*. New York: Grosset & Dunlap.

Derrida, Jacques. 1978. *Writing and Difference*. Trans. A. Bass. Chicago: University of Chicago Press.

Douglas, Frederick. 1855. *My Bondage and My Freedom*. New York and Auburn: Miller, Orton and Mulligan.

Drinnon, Richard. 1980. *Facing West: The Metaphysics of Indian-Hating and Empire-Building*. New York: Meridian.

Du Bois, W. E. B. 1921. *Darkwater; Voices from within the Veil*. New York: Harcourt Brace and Co. Reprint, New York: Schocken Books, 1969.

———. 1940. *Dusk of Dawn: An Essay Toward an Autobiography of a Race Concept*. Reprint, New York: Schoken Books, 1968.

———. 1961. *The Souls of Black Folk*. New York: Fawcett Publications, Inc.

———. 1971. *The Seventh Son: The Thought and Writings of W. E. B. DuBois, Vol. 1*. Ed. Julius Lester. New York: Vintage Books.

———. 1982. *Writings by W. E. B. DuBois in Periodicals Edited by Others, Vol. 2*. Comp. and ed. H. Aptheker. Millwood, New York: Kraus-Thomson Organization Limited.

———. 1997 (new edition). *John Brown: A Biography*. Primary documents and intro. by John David Smith. Armonk, NY and London: M. E. Sharpe.

Dumm, Thomas. 1993. "The New Enclosures: Racism in the Normalized Community." *Reading Rodney King, ReadingUrban Uprising*. Ed. R. Gooding-Williams. New York: Routledge, 178–195.

Dussel, Enrique. 1980. "Christian Art of the Oppressed in Latin America (Towards an Aesthetics of Liberation)." In *Symbol and Art in Worship*. Ed. Luis Maldonado and David Power. New York: Seabury Press, 40–52.

———. 1995. *The Invention of the Americas: Eclipse of "the Other" and the Myth of Modernity*. New York: Continnum.

Dyer, Richard. 1988. "White." *Screen* 29/4: 44–64.

Earl, Riggins Jr. 1993. *Dark Symbols, Obscure Signs: God, Self, and Community in the Slave Mind*. Maryknoll: Orbis Press.

Fanon, Frantz. 1967. *Black Skins, White Masks*. Trans. C. L. Markmann. New York: Grove Press. Reprint, New York: Grove Weidenfeld, 1991.

Feldman, Allen. 1991. *Formations of Violence: The Narrative of the Body and Political Terror in Northern Ireland*. Chicago: the University of Chicago Press.

Fergusson, Fergusson, Francis. 1989. *The Idea of the Theater*. Princeton: Princeton University Press.

Foucault, Michel. 1970. *The Order of Things*. London, Tavistock; New York: Pantheon.

———. 1977. *Discipline and Punish: The Birth of the Prison*. New York: Pantheon Books.

———. 1980. *Power/Knowledge: Selected Interviews and Other Writings, 1972–1977*. Ed. C. Gordon. Trans. C. Gordon et al. New York: Pantheon Books.

———. 1984. "Truth and Power." In *Foucault Reader*. Ed. Paul Rabinow. New York: Pantheon Books, 51–75.

Frankenberg, Ruth. 1993. *The Social Construction of Whiteness: White Women, Race Matters*. Minneapolis: University of Minnesota Press.

Fraser, Nancy. 1989. *Unruly Practices: Power, Discourse and Gender in Contemporary Social Theory*. Minneapolis: University of Minnesota Press.

Gates, Henry Louis, Jr. 1985. "Editor's Introduction: Writing 'Race' and the Difference It Makes." In *"Race," Writing and Difference*. Chicago: University of Chicago Press, 1–20.

———. 1988. *The Signifying Monkey: A Theory of Afro-American Literary Criticism*. New York: Oxford University Press.

Gilmore, Ruth. 1993. "Terror Austerity Race Gender Excess Theater." In *Reading Rodney King, Reading Urban Uprising*. Ed. R. Gooding-Williams. New York: Routledge, 23–37.

Gilroy, Paul. 1987. *There Ain't No Black in the Union Jack: The Cultural Politics of Race and Nation*. London: Melbourne: Hutchinson. Reprint, Chicago: University of Chicago Press, 1991.

———. 1993. *The Black Atlantic: Modernity and Double Consciousness*. Cambridge: Harvard University Press.

Girard, Rene. 1977. *Violence and the Sacred*. Trans. P. Gregory. Baltimore: The Johns Hopkins University Press.

Grant, Jacquelyn. 1993. "The Sin of Servanthood: And the Deliverance of Discipleship." In *A Troubling in My Soul: Womanist Perspectives on Evil and Suffering*. Ed. E. Townes. Maryknoll: Orbis Books.

Gutierrez, Gustavo. 1973. *A Theology of Liberation: History, Politics, and Salvation*. Trans. and ed. Sister C. Inda and J. Eagleson. Maryknoll: Orbis Books.

Habermas, Jurgen. 1987. *The Philosophical Discourse of Modernity: Twelve Lectures*. Trans. F. Lawrence. Cambridge: MIT Press.

Hall, Stuart. 1992. "What is This 'Black' in Black Popular Culture?" In *Black Popular Culture*. Ed. G. Dent. Seattle: Bay Press, 21–33.

Harding, Vincent. 1979. "Black Power and the American Christ." In *Black Theology: A Documentary History, 1966–1979*. Ed. Gayraud S. Wilmore and James H. Cone. Maryknoll: Orbis Books, 35–42.

Harris, Cheryl. 1993. "Whiteness as Property." *Harvard Law Review* 106/8: 1709–1791.

Hartigan, John, Jr. 1999. *Racial Situations: Class Predicaments of Whiteness in Detroit*. Princeton: Princeton University Press.

Haymes, Stephen N. 1995. *Race, Culture, and the City: A Pedagogy for Black Urban Struggle*. Albany: State University of New York Press.

Hegel, G. W. F. 1977. *Phenomenology of Spirit*. Trans. A. V. Miller; analysis and foreword by J. N. Findley. Oxford: Clarendon Press.

Herrnstein, Richard J. and Murray, Charles. 1994. *The Bell Curve: Intelligence and Class Structure in American Life*. New York: Free Press.

Holt, Thomas. 1990. "The Political Uses of Alienation: W. E. B. DuBois on Politics, Race, and Culture, 1903–1940." *American Quarterly Review* 42/2: 301–323.

hooks, bell. 1992. *Black Looks: Race and Representation*. Boston: South End Press.

hooks, bell and Cornel West. 1991. *Breaking Bread: Insurgent Black Intellectual Life*. Boston: South End Press.

Hopkins, Dwight. 1993. *Shoes That Fit Our Feet: Sources for a Constructive Black Theology*. Maryknoll: Orbis Books.

———. 1997. "Postmodernity, Black Theology of Liberation and the U.S.A.: Michel Foucault and James H. Cone." *Liberation Theologies, Postmodernity, and the Americas*. New York: Routledge, 205–221.

Hudson, Michael. 1996. *Merchants of Misery: How Corporate America Profits From Poverty*. Monroe, Maine: Common Courage Press.

Hurston, Zora Neal. 1969. *Dust Tracks on the Road*. New York: Arno Press and the New York Times.

Ignatiev, Noel. 1995. *How the Irish Became White*. New York: Routledge.

Isasi-Diaz, Ada Marie. 1996. *Mujerista Theology: A Theology for the 21st Century*. Westminster: John Knox Press.

James, Henry Sr. 1879. *Society: The Redeemed Form of Man*. Cambridge: Hougton, Osgood & Co., 44–49.

James, William. 1990. *The Varieties of Religious Experience*. New York: Vintage Books/Library of America.

Jordan, Winthrop D. 1968. *White over Black: American Attitudes toward the Negro, 1550–1812*. New York: W. W. Norton & Co.

Kaiser, Rudolf. 1991. *The Voice of the Great Spirit: Prophecies of the Hopi Indian*. Trans. W. Wunsche. Boston: Shambhala.

Kant, Immanuel. 1956. *Critique of Practical Reason*. Trans. L. W. Beck. New York: MacMillan.

Kearney, Richard. 1988. *Transitions: Narratives in Modern Irish Culture*. Manchester, UK: Manchester University Press; New York, NY, USA: Distributed by St. Martin's Press.

Kinukawa, Hisako. 1994. *Women and Jesus in Mark: A Japanese Feminist Perspective*. Maryknoll: Orbis.

Klor de Alva, Jorge. 1996. "Is affirmative Action a Christian Heresy?" *Representations* 55 (Summer 1996), 59–73.

Klor de Alva, Jorge, Shorris, Earl, and West, Cornel. 1996. "Our Next Race Question: The Uneasiness Between Blacks and Latinor," *Harper's Magazine* (April 1996): 55–63.

Kochman, Thomas. 1981. *Black and White Styles in Conflict*. Chicago: University of Chicago Press.

Kwok, Pui-lan. 1993. "Racism and Ethnocentrism in Feminist Biblical Interpretation." In *Searching the Scriptures: A Feminist Introduction*. New York: Crossroad.

Laclau, Ernesto and Mouffe, Chantal. 1985. *Hegemony and Socialist Strategy: Towards a Radical Democratic Politics*. Trans. W. Moore and P. Cammack. London: Verso.

Lattany, Kristin Hunter. 1994 (1993). " 'Off-timing': Stepping to the Different Drummer." *Lure and Loathing: Essays on Race, Identity, and the Ambivalence of Assimilation*. Ed. Gerald Early. New York: Penguin Books, 163–174.

Lefort, Claude. 1986. *The Political Forms of Modern Society: Bureaucracy, Democracy, and Totalitarianism*. Ed. J. B. Thompson. Cambridge, Mass: MIT Press.

Leki, Peter. 1996. "Why Be White?" *The Reader*, June 28, 1996, section 1, 12–17.

Long, Charles. 1985. "A Look at the Chicago Tradition in the History of Religions: Retrospect and Future." In *The History of Religions: Retrospect and Prospect*. Ed. Joseph Kitagawa.New York: MacMillan, 87–104.

———. 1986. *Significations: Signs, Symbols, and Images in the Interpretation of Religion*. Philadelphia: Fortress Press.

Madhubuti, Haki R. 1993. "Introduction; Same Song, Different Rhythm." In *Why L.A. Happened: Implications of the '92 Los Angeles Rebellion*. Ed. H. Madhubuti. Chicago: Third World Press, xiii–xvii.

Malcolm X. 1964. *The Autobiography of Malcolm*. Originally published by Malcolm X and Alex Haley. Published, New York: Ballantine, 1973.

Mandelbaum, Allen. 1980. Introduction, *The Divine Comedy of Dante Alighieri, Inferno*. Trans. Allen Mandelbaum, notes by Mandelbaum and Gabriel Marruzzo, with Laury Magnus. Berkeley: University of California Press.

Marx, Karl. 1967. *Capital: A Critique of Political Economy, Vol. 1*. Ed. F. Engels. Trans. S. Moore and E. Aveling. New York: International Publishers.

Masciandaro, Franco. 1991. *Dante as Dramatist: The Myth of the Earthly Paradise and Tragic Vision in the Divine Comedy*. Philadelphia: University of Pennsylvania Press.

McDermott, John J. (ed.). 1968. *The Writings of William James*. New York: Modern Library.

Menocal, Maria Rosa. 2002. *The Ornament of the World: How Muslims, Jews, and Christians Created a Culture of Tolerance in Medieval Spain*. Boston: Little Brown and Co.

Miller, Perry. 1956. *Errand Into the Wilderness*. Cambridge: Harvard University Press.

Mills, Charles. 1997. *The Racial Contract*. Ithaca and London: Cornell University Press.

Morrison, Toni. 1970. *The Bluest Eye*. New York: Washington Square Press.

———. 1987. *Beloved: A Novel*. New York: Knopf: Distributed by Random House.

Mudimbe, V. Y. 1988. *The Invention of Africa: Gnosis, Philosophy, and the Order of Knowledge*. Bloomington: Indiana University Press.

Nietzsche, F. 1990. *Twilight of the Idols*. New York: Penguin Books.

Omi, Michael and Winant, Howard.1994. *Racial Formation in the United States From the 1960s to the 1990s*. 2nd Edition. New York: Routledge.

Otto, Rudolph. 1950. *The Idea of the Holy*. Trans. John W. Harvey. London: Oxford University Press.

Ozment, Steven E. 1971. *The Reformation in Medieval Perspective*. Chicago: Quadrangle Books.

Pagden, Anthony. 1995. *Lords of All the World: Ideologies of Empire in Spain, Britain, and France, c. 1500–c. 1800*. New Haven: Yale University Press.

Parker, Sherry. 1999. "Making a Case For Personal Social Security Accounts." *Self-Employed America* (Nov.–Dec. 1999): 15.

Perkinson, James. 1994. "On Being 'Doubled': Soteriology at the White End of Black Signifyin(g)." *Koinonia Journal* (Fall, 1994): 176–205.

———. 1997. "Beyond Occasional Whiteness." *Cross Currents* Vol. 47, No. 2 (July 1997): 195–209.

———. 1998. "A Socio-Reading of the Kierkegaardian Self: or, the Space of Lowliness in the Time of the Disciple." In *Kierkegaard: The Self in Society*. London: McMillan Press, 1998, 156–172.

———. 1999. "Rage With a Purpose, Weep Without Regret: A White Theology of Solidarity." *Soundings* Vol. 82, Nos. 3–4 (Fall/Winter 1999): 437–463.

———. 2000. "The Color of My Enemy: Comparative Racializations of the Absolute Other in Black and White Theologies." *Cross Currents* Vol. 50, No. 3 (Fall, 2000): 349–368.

———. 2001. "Theology and the City: Learning to Cry, Struggling to See." *Cross Currents* Vol. 51, No. 3 (Spring 2001): 95–114.

———. 2002. "The Body of White Space: Beyond Stiff Voices, Flaccid Feelings, and Silent Cells." In *Revealing Male Bodies*. Bloomington, Indiana: Indiana University Press, 173–197.

———. 2003, "Black Theology and the White Church in the Third Millennium: Like A Thief in the Night." In *Living Stones in the Household of God: The Legacy and Future of Black Theology*, ed. L. Thomas, Minneapolis: Augsburg Fortress, 83–98.

———. 2004. "White Church Response to Black Theology: Like a Thief in the Night." *Theology Today* Vol. 60, No. 4 (Jan. 2004), 508–524.

Perry, Benita. 1987. "Problems in Current Theories of Colonial Discourse." *Oxford Literary Review* 9: 27–58.

Pieris, Aloyius. 1988. *An Asian Theology of Liberation*. Maryknoll: Orbis.

Pike, Frederick B. 1995 (1990). "Latin America." *The Oxford Illustrated History of Christianity*. Ed. John McManners. Oxford: Oxford University Press, 420–454.

Porterfield, Amanda, 1987. "Shamanism: A Psychosocial Definition." *Journal of the American Academy of Religion* 55 (1987): 725–729.

Quinn, Daniel. 1992. *Ishmael*. New York: Bantam Books.

Rabasa, Jose. 1993. *Inventing America: Spanish Historiography and the Formation of Eurocentrism*. Norman, Okla.: University of Oklahoma Press.

Raboteau, Albert. 1978. *Slave Religion: The "Invisible Institution" in the Antebellum South*. New York: Oxford University Press.

Roediger, David. 1991. *The Wages of Whiteness: Race and the Making of the American Working Class*. London; New York: Verso.

Rifkin, Jeremy. 1995. *The End of Work: The Decline of the Global Labor Force and the Dawn of the Post-Market Era*. New York: G. P. Putnam's Sons.

Rose, Tricia. 1994. *Black Noise: Rap Music and Black Culture in Contemporary America*. Hanover, NH: Wesleyan University Press, Published by University Press, of New England.

Ruether, Rosemary Radford. 1983. *Sexism and God Talk: Toward a Critical Feminist Theology*. Boston: Beacon Press.

Schlosser, Eric. 1998. "The Prison Industrial Complex." *The Atlantic Monthly* (December 1998): 51–79.

Scott, James. 1990. *Domination and the Arts of Resistance: Hidden Transcripts*. New Haven: Yale University Press.

Segundo, Juan Luis. 1984. *Faith and Ideologies*. Trans. J. Drury. Maryknoll: Orbis Books.

Shorris, Earl. 1996. "Our Next Race Question: The Uneasiness Between Blacks and Latinos." *Harper's Magazine*, April, 56–63.

Slaughter, Thomas F. Jr. 1983. "Epidermalizing the World: A Basic Mode of Being Black." *Philosophy Born of Struggle: Anthology of Afro-American Philosophy from 1917*. Ed. L. Harris. Dubuque: Kendall/Hunt Publishing Co., 283–285.

Song, Choan-Seng. 1982. *The Compassionate God*. Maryknoll: Orbis.

Smith, Theophilus. 1989. "The Spirituality of Afro-American Traditions." *Christian Spirituality: Post-Reformation and Modern*. Ed. Louis Dupre' and E. Saliers New York: Crossroads, 372–414.

———. 1994. *Conjuring Culture: Biblical Formations of Black America*. New York: Oxford University Press.

Smitherman, Geneva. 1994. *Black Talk: Words and Phrases from the Hood to the Amen Corner*. New York: Houghton Miflin.

Spivak, Gayatri. 1988. "Can the Subaltern Speak." *Marxism and the Interpretation of Culture*. Ed. C. Nelson and L. Grossberg. Chicago: University of Illinois Press, 271–313.

———. 1990. *The Postcolonial Critic: Interviews, Strategies, Dialogues*. Ed. S. Harasym. New York: Routledge.

———. 1992. "The Politics of Translation." In *Destabilizing Theory: Contemporary Feminist Debates*. Ed. M. Barrett and A. Phillips. Stanford: Stanford University Press, 177–200.

Stringfellow, William. 1973. *An Ethic for Christians and Other Aliens in a Strange Land*. Waco, Texas: Word Books.

Stroupe, Nibs and Fleming, Inez. 1995. *While We Run this Race: Confronting the Power of Racism in a Southern Church*. Maryknoll: Orbis Books.

Swanson, Tod. 1994. "Colonial Violence and Inca Analogies to Christianity." In *Curing Violence: Essays on Rene Girard* Ed. M. I. Wallace and T. Smith. Sonoma, C: Polebridge Press.

Thistlethwaite, Susan Brooks. 1989. *Sex, Race, and God: Christian Feminism in Black and White*. New York: Crossroad.

Todorov, Tzvetan. 1983. *The Conquest of America: The Question of the Other*. Trans. R. Howard. New York: Harper & Row.

Van der Leeuw, Gerardus. 1938. *Religion in Essence and Manifestation*. Trans. J. E. Turner. London: George Allen & Unwin.

Warrior, Robert Allen. 1989. "Cananites, Cowboys, and Indians: Deliverance, Conquest, and Liberation Theology Today." *Christianity and Crisis* 49 (September 11): 261–265.

Weber, Max. 1958. *The Protestant Ethic and the Spirit of Capitalism*. Trans. T. Parsons. New York: Charles Scribner.

Welch, Sharon D. 1985. *Communities of Resistance and Solidarity: A Feminist Theology of Liberation*. Maryknoll, NY: Orbis Press.

Wessels, Anton. 1990 (1986). *Images of Jesus: How Jesus is Perceived and Portrayed in Non-European Cultures*. Grand Rapids: Eerdmanns Publishing Co.

West, Cornel. 1982. *Prophesy Deliverance: An African-American Revolutionary Christianity*. Philadelphia: Westminster Press.

———. 1989a. *The American Evasion of Philosophy: A Genealogy of Pragmatism*. Madison, Wis.: University of Wisconsin Press.

————. 1989b. "Black Culture and Postmodernism." In *Remaking History*. Ed. B. Kruger and P. Mariani. Seattle: Bay Press, 87–96.

————. 1991. "The New Cultural Politics of Difference." In *Out There: Marginalization and Contemporary Cultures*. Ed. R. Ferguson, M. Gever, T. Minh-ha, and C. West. Cambridge: MIT Press, 19–38.

White, Hayden. 1978. *Tropics of Discourse: Essays in Cultural Criticism*. Baltimore: Johns Hopkins University Press.

Williams, Delores. 1993. *Sisters in the Wilderness: The Challenge of Womanist God-Talk*. Maryknoll: Orbis Books.

Williams, Patricia. 1991. *The Alchemy of Race and Rights*. Cambridge: Harvard University Press.

Willis, Susan B. 1987. *Specifying: Black Women Writing the American Experience*. Madison, Wis.: University of Wisconsin Press.

Wilmore, Gayraud S. 1974. "Black Theology." *International Review of Missions* 63: 227–239.

Wink, Walter. 1984. *Naming the Powers: The Language of Power in the New Testament*. Minneapolis: Fortress Press.

————. 1986. *Unmasking The Powers: the Invisible Forces That Determine Human Existence*. Minneapolis: Fortress Press.

————. 1992. *Engaging the Powers: Discernment and Resistance in a World of Domination*. Minneapolis: Fortress Press.

Witvliet, Theo. 1985. *A Place in the Sun*. Maryknoll: Orbis.

————. 1987. *The Way of the Black Messiah: The Hermeneutical Challenge of Black Theology as a Theology of Liberation*. London: SCM Press.

Wright, Richard. 1937. *Black Boy: A Record of Childhood and Youth*. Originally published by R. Wright. Published, New York: Harper & Row, 1945. Reprint, 1964.

————. 1940. *Native Son*. New York: Harper.

Index